The Digital Turn: User's Practices and Cultural Transformations

Pille Runnel / Pille Pruulmann-Vengerfeldt /
Piret Viires / Marin Laak (eds.)

The Digital Turn: User's Practices and Cultural Transformations

PL ACADEMIC RESEARCH

Bibliographic Information published by the Deutsche Nationalbibliothek
The Deutsche Nationalbibliothek lists this publication in the Deutsche Nationalbibliografie; detailed bibliographic data is available in the internet at http://dnb.d-nb.de.

English language editor: Daniel Edward Allen

Copy-editors: Toivo Sikka, Tuuli Kaalep

Layout: Krete Pajo

Cover photo:
Leonhard Lapin, „Cosmic Machine"

This book was published with the support
of Estonian Science Foundation grants no. 7679, no. 8006
and the target financed research project no. SF0030065s08.

Library of Congress Cataloging-in-Publication Data

The digital turn : user's practices and cultural transformations /
Pille Runnel, Pille Pruulmann-Vengerfeldt, Piret Viires, Marin Laak (eds.).
pages cm
ISBN 978-3-631-64035-7
1. Information technology—Social aspects. 2. Mass media—Technological innovations. 3. Digital media—Social aspects. I. Runnel, Pille. II. Pruulmann-Vengerfeldt, Pille. III. Viires, Piret, 1963-
HM851.D54525 2013
302.23'1—dc23

2013008646

ISBN 978-3-631-64035-7

PL Academic Research is an Imprint of Peter Lang GmbH

Peter Lang – Frankfurt am Main · Berlin · Bruxelles · New York ·
Oxford · Warszawa · Wien

www.peterlang.de

Table of Contents

The Challenge of the Digital Turn

Pille Pruulmann-Vengerfeldt, Pille Runnel, Marin Laak, Piret Viires

The past fifty years or so have seemingly been a whirlwind of turns. There have been discussions in the social sciences and humanities about the linguistic turn, the cultural turn, the pictorial turn, the cognitive or performative turn. Doris Bachmann-Medick (2006) has found that the cultural turn can be divided into seven distinct turns: the interpretative turn, the performative turn, the reflexive/literary turn, the postcolonial turn, the translational turn, the spatial turn and the iconic turn. Most of them can be considered turns in scientific rhetoric and the apparatus of understanding, expressing new, transdisciplinary approaches which have enabled reconstruction of the objects of research and discovery of new ones. However, this book looks at the digital turn, in which, in addition to the changes in scholarship, 'digital' definitely also has a material and formal aspect to it, a significant shift not only in the forms, environments and technologies, but also in the much deeper influence on the socio-cultural relations and interactions that these new forms and environments support and foster.

The book you are currently holding has two sections: User's Practices and Cultural Transformations. The aim of the book is to discuss how the digital turn in the cultural field has resulted in increasing attention being paid to users and their practices of consuming and creating digital content. At the same time this has resulted in some remarkable transformations both in cultural institutions as well as the forms and modes of cultural content.

The digital turn hereby implies that changes in the use and application of digital technology bring on changes in practice and in the relationships between cultural institutions and audiences. We approach the changes in society from the structural (institutional) as well as from the agential (audiences, users, individuals) perspective. Although it is clear that the rising importance of the digital or the new media influences cultural representations as well as forms of cultural participation and socialisation, the digital turn does not mean turning away or turning to a new direction of culture. The authors represented in this book share the view that there is no fear of new media pushing aside traditional cultural forms, acknowledging at the same time that the scope of this cultural change, involving both the digital and non-digital, is still far from understood (Lauristin). By offering their insights into this question, many of the authors in this book argue that the digital turn can be conceptualised only if seen as a part of the wider dynamic – the turn incorporates both digital and non-digital aspects of culture. Instead of celebrating a digital revolution we argue that what we are seeing are evolutions within the cultural processes where the digital is only a part of the overall change.

The book brings together components of the classic model of text, producer and reader (Hall), where the practices embody the reading and producing of cultural heritage. These practices are seen as being on the intersection between individuals and structure, the embodied sets of activities that humans perform with varying degrees of regularity, competence and flair (Postill). In this book, the institutional practices and user (visitor, audience, author, etc.) practices meet.

User has a very open meaning in this book. On the one hand, users are 'people who use' the (digital) content of the heritage institutions, on the other hand, the usage itself is seen not as a passive consumption, but it has a distinct part in active meaning-making, production and participation.

The heritage institutions themselves can be seen as active users of existing digital environments (e.g. Facebook in the chapter by Schick and Damjkaer), creators of digital environments (e.g. Laak) as well as participants in cultural creation (Pruulmann-Vengerfeldt and Runnel). This points to the blurring boundaries of users and producers summarised often in terms 'produsers' or 'prosumers'. Users have the capacity to change existing understandings of cultural resources by adding layers of information to institutional representations (Casado-Neira) and challenge the public and the private through personal identity management, simultaneously flaunting and hiding information (Koosel). While several of the authors in this book discuss how the individual meets the institutional, Bannier and Vleugels look, in their study, at how in the framework of Web 3.0, the cultural content generated by institutions and users collides with a third kind of cultural content, generated by machines, an opportunity offered by the development of a hybrid, semantic and intelligent web.

These examples also point to the blurring boundaries of the institutional and individual, the contestations of the structure and agency dichotomy as the institutions become agents and individual agents become parts of the structural fabric of cultural institutions.

Despite this we must not forget that while a multitude of examples support the understanding of the user per se as a participant in cultural processes, digital heritage does not automatically lead to an interactive and participatory culture, as Mostmans and van Passel show. This book offers critical insights into the user as a participant by questioning, especially in the context of cultural heritage institutions, how much the availability of heritage content actually triggers or facilitates participation. Sometimes, the usage of cultural content is still limited by basic questions of access, and as Weisen shows in his article about the accessibility of digital culture for disabled people, the issue of access is still far from being solved.

From the memory institutions' perspective, the digital turn is closely connected to the representation of current cultural heritage. New research looks at the issues of digital memory, approaching the new media environment as a memory space. This aspect of memory and remembrance brings an additional, diachronic, dimension to storytelling and new forms of storytelling emerge in literary and cultural historiography (Laak).

Different chapters of the book discuss the transformation of culture through a variety of practices on the one hand made possible through digital technologies, and on the other hand, brought to light and made visible because of other changes in society. Thus many chapters discuss the possibilities of participation within these cultural institutions not only enabled by digital technologies, but as an indication that participation is an important issue also without technological components (Carpentier; Olsson and Svensson) and can happen in different ways. As a part of the same discussion, Kaun and Östman direct attention to the idea of the playful and fun as a way to foster engagement and belonging.

In this book, the archives, libraries and museums meet authors, individuals, tourists. The sociology and media studies perspectives meet the historical, literary and philosophical traditions. The aim of this diversity is to bring together different perspectives between the same covers, to share the positives of the interdisciplinary approach and to bring attention to the diversity of the field. The book takes the theoretical perspectives and examples of good practices and translates them into some universal ideas. At the same time, the forever-questioning nature of research translates into critical and analytical accounts of these practices. It keeps on asking whether the promises of new forms of culture, new online environments and changes in production and reception practices are significantly new. As the chapters in this book point out, different institutions are facing similar struggles, and not only can the best practices for the museums be found in other museums, but as Olsson and Svensson demonstrate we need to learn from others as well. They use the example of Moderskeppet, a web company representing the commercial sphere, analysing it as the best web practice to highlight the importance of interactivity and public contribution.

The discussions about the digital have mainly focused on the influence of technologies on the content – the artefacts, the texts and materials that have been made available online. In this book literary scholar Raine Koskimaa highlights how the invention of hypertext has given rise to a changed conceptualisation of the text itself. This, he argues, makes cultural logic gradually give way to the new logic of Castells's "Internet Galaxy". Koskimaa is supported by Markku Eskelinen, who moves away from the old cultural logics (represented by traditional theories of inter- and transtextuality) and brings to our attention the new types of relations between and within texts as well as theorising them from a cybertextual perspective.

We feel that the digital and overall participatory transformations within society have called for increasing attention to the people. from the people. Often the new aspects of the digital technologies are seen to be inherently social. The new Web 2.0 is conceptualised through contributory and participatory aspects of cultural production. However, in this book, in the context of literature and cyberspace, Viires and Sarapik bring to our attention the fact that the creation of digital text can still be analysed from the perspective of essentially individual and solitary processes. This is important despite the increasing collectivism and the shared authorship and reader-viewer interactions that are increasingly a part of the text creation process. In addition, Beyl explores how the writer appropriates weblogs as a communicative medium, to explore as a tool of self-conceptualisation and as a tool with which to (de)mystify the author's aesthetic role and position in society.

The changing practices of cultural production and consumption have also meant that many cultural institutions feel the need to reinstate their positions within the structure of the society. What is the role of the library, when every book is digitally available for download? What is the role and practice of archives, when digital technologies seem to enable endless storage capacities? What is the role of museums in a world where the aura of the original seems to be under siege? Who is included and excluded in the practices of culture making, who is the author (Viires and Sarapik)? Or is the author long dead as proclaimed by Barthes (Carpentier)? When the world has taken major turns in the past twenty-something years, are cultural institutions the places for learning for the future or should the musty old cabinets be locked and sealed for future generations? Many of these questions are related to the digital turn, and they are discussed in the different chapters of this book. For instance, Lepik discusses how calling people users, visitors, stakeholders or audiences will make the relationship significant and, to a certain extent, re-instate the position of the institution. Carpentier looks at the history of cultural participation to argue that the seemingly novel and unique changes in power relations proclaimed by the "death of the author" slogan have been a part of the cultural process throughout the 20th century. However, this still does not mean that discursive structures as the conditions of possibility for the organisation of participation are simple or solved.

The digitisation of heritage is still too often seen as an aim on its own because the technology seemingly enables wider access and better availability of materials. Volt and Andresoo, and Weisen, among others, discuss the variety of challenges posed by digitisation relating to issues of access, long term preservation, formats, etc. The book as a whole, however, attempts to bring the discussion towards the wider cultural implications of these challenges. The digital turn has resulted in different discussions across paradigms and focuses; the idea of the book is to bring together some of these discussions to learn from each other. Too often in the practice of heritage institutions the digital turn is seen as an aim on its own: the digitisation process is seen as a mediation of culture and a change of format as the only way to promote culture. In this approach, the digital turn is separated from the rest of the cultural process, seen as being over as soon as cultural content has been made available to the public via the digital. It also denies the self-critical awareness of the cultural institutions and other related stakeholders and sees them as one among the multitude of agents in a wider socio-cultural process, involving both the online and offline. Digital is often still a destination point. In this book the authors aim to challenge this view. In the closing chapter, Farouk Seif argues that technology is not and should not be the destination of our making. Rather, we should be looking forward to a future that transcends virtual reality and incorporates technical and technological aspects of culture as a part of the authentic experience.

Thus, this book looks at the range of different aspects of interrelations between technology and culture – but can we truly claim that this is the digital turn that we are facing? Many would see that the explosive growth of new technologies and platforms, processes speeding up and the increasing number of users are enough to justify the talk about the digital turn. The facts that these environments are increasingly considered normal and that our everyday lives are saturated with technologies are remarkable, but we think that they are still not a reason enough to talk about the digital turn. For us and for many of the authors in this book, the digital turn means a variety of complex changes in interaction with each other in our social and cultural environment. It means that the user is much more central to cultural processes, and the cultural elites have to consider a more diverse range of players in culture-making processes. The digital technologies make the public, the audiences and the users more visible. These interactions and produsages do bring new cultural forms, but as many of the texts indicate, they do not replace the old forms. Therefore we see that such a transdisciplinary book as this one, where the humanitarian and social sciences meet, provides a unique perspective on investigating the cultural processes in which the digital focuses the attention of the user on the collective and multi-party aspects of culture-making and consumption. The turn to the digital changes not only cultural forms and interactions, but also the institutions and their relationships with audiences. Therefore the two sections of this book, user practices and cultural transformations, are interdependent and can thus only shed adequate light on the true nature of the digital turn together.

Acknowledgements

The publication of this book was made possible by the Estonian Science Foundation grants no. ETF7162, ETF7679 and ETF8006, the target financed research projects no. SF0180017s07, SF0180002s07, SF0030065s08, SF0030054s08 and the Estonian National Museum.

References

Bachmann-Medick, Doris. *Cultural Turns: Neuorientierungen in den Kulturwissenschaften.* Hamburg: Rowohlt, 2006. Print.

Hall, Stuart. "Encoding/decoding." *Culture, Media, Language: Working Papers in Cultural Studies.* London: Hutchinson, 1980. 117-127. Print.

Postill, John. "Introduction: Theorising Media and Practice." *Theorising Media and Practice.* Eds. Birgit Bräuchler and John Postill. Berghahn, 2010. 1-33. Print.

Orienting the Heritage Institution towards Participatory Users in the Internet

Pille Pruulmann-Vengerfeldt

Pille Runnel

Agnes Aljas

University of Tartu
Estonian National Museum
Veski 32
51014 Tartu, Estonia
Pille.Vengerfeldt@ut.ee
Pille.Runnel@erm.ee
Agnes.Aljas@ut.ee

1. Introduction

This paper looks at different experiences Estonian heritage institutions have in recruiting active participants in their work. The focus is mainly on providing services through online channels and including active participants in the work on museums, archives and libraries. We base our discussions on the work done in two research projects: "Developing Museum Communication in the 21st Century", and "The Problems of Transformation and Reception of Cultural Heritage in the Digital Age", which both look at the changing relations between audiences and heritage institutions. We will base our discussion on the empirical case studies, using four different heritage institutions in Estonia as examples and as the basis of analysis – the Estonian National Museum, the Estonian Literary Museum, the National Archives of Estonia and University of Tartu Library. We have combined several research methods and looked at several types of respondents over different points in time. The paper aims to be a more reflexive overview of audience relations and participation in heritage institutions.

In using the notions of 'audiences' and 'users' interchangeably, we assume that audiences are active despite the communication channels used. In increasing use of ICT solutions in heritage institutions, audiences are assigned the even more active role of 'users' or 'produsers', as used by Axel Bruns. Coming from a media studies perspective we approach the audiences in the heritage institutions from these angles. We place this article at the crossroads of different disciplines in the hope of adding value to the practical applications on which we are working.

2. Active audiences and heritage institutions

Active audiences and participation are not entirely new phenomena in the context of heritage institutions (see also Carpentier). Many museums have built their collections, i.e. their specific body of knowledge, using objects and information sourced from the people. Archives depend on the public for the provision of documents and libraries use groups of readers or reader statistics to help formulate their collection policies. However, in all of these cases the heritage worker plays the role of gate-keeper, moderating and limiting the participation for particular purposes.

This expert or gate-keeper position originates from and at the same time also produces a way of understanding of the concept of 'heritage', which implies that cultural heritage is 'real'. Kristin Kuutma, Estonian researcher of heritage politics, who analyses how the notion of heritage is constructed, states that

> [c]ultural heritage becomes real when someone identifies it as such, which denotes a process of knowledge production that involves academic research. ... The awareness of heritage is epistemologically related to scholarship of history, art, ethnology, folklore, etc. – heritage is a certain way of knowing cultural objects, sites or practices (7-8).

Heritage is thus related to expertise and symbolic power.

According to both public understanding and the cultural heritage politics framework, the maintenance and safeguarding of the perceived common heritage has belonged to the realm of the heritage institutions. Through the role appropriated by this approach to cultural heritage, these institutions have also been instruments in the development of a sense of collective identity and citizenship and are thus arenas for the distribution of socio-political capital.

In the current article we look at this set of issues from one particular perspective, asking how and why this perceived role of heritage guardians can also serve as an obstacle or barrier to involving people to the field of the heritage in contemporary societies. More widely, this discussion is related to questions about whom does heritage empower and the question of whether it is possible to use the Internet to attract wider audiences to be engaged in the use and creation of heritage.

Our analysis has indicated that heritage institutions have a perceived expert position, which can in the current situation, when technology permits open and wide participation at low costs, become an invisible barrier to audience participation. The Internet seems to provide ample opportunities to engage the public in a dialogue with the heritage institutions; however, two-way communication assumes not only the existence of a communication channel, but also willing parties who are interested in communication. We claim that here the way in which heritage institutions are perceived, and act, as gatekeepers may be part of the reason why audiences/users would not be that keen on participating and contributing. We are not arguing that this role should be left behind as many researchers have also shown the audience or users would still like to be distant from key decision-making and leave it to the experts (see, for example, Davies). Rather, the challenge is in empowering audiences, the prerequisite of which is an acknowledgement of the barrier between the two by the heritage institutions and their active search for ways to overcome it.

Next, we will have a look at Internet users in order to understand better different potential users of the applications provided by the heritage institutions. We will critically examine the supposedly active user, at whom many of the content creation applications are aimed. We argue that in order to activate participation in heritage institutions, people should not be left to find their way around, rather this needs to be a conscious attempt from the heritage institutions to create participation spaces that are user friendly and engaging.

3. Internet users and uses

Together with the increase in Internet users and the rise of Web 2.0 (O'Reilly) the arrival of active users has been hailed. The Internet today enables the use of more and more user-friendly technologies in order to actively foster participation in creating and providing content online. Web 2.0 applications such as blogs, social networking sites, and photo and video communities provide increasing opportunities for everyone to become their own publisher and have visibility in the online environment. However, not everyone wants to put him or herself online. When analysing the largest video-based user-generated content environment, YouTube, Cha et al. point out that in general users of this environment are rather passive in using Web 2.0 features like commenting or rating the content.

Jakob Nielsen has made a famous observation on participatory content online stating that in most online communities ninety percent of users are lurkers who never contribute, nine percent contribute a little and one percent account for almost all the action. Similar tendencies can be seen in Estonia where by the end of 2008, seventy percent of Estonians used the Internet and of them only 38% had only ever tried to upload a photo – the most common content production activity of all – while only ten percent had ever commented on an online news item (Meema). The challenge here is to forgo the hype of participatory media and look deeper at the ways and styles of participation. Active audiences in the Internet are first and foremost either seeking information necessary for their work or personal lives or communicating with friends and peers for leisure and entertainment purposes. Because outside the Internet museums, libraries and archives play only a small part in an individual's life, expecting a dominant role for them in the online environment is unrealistic. However, one has to keep in mind that providing online access to heritage materials or opening the collections to participation, work can be done in order to locate the most appropriate target group for whom some kind of heritage related application might be the most important Internet application ever.

In a number of previous studies (see for instance Keller et al.; Laak; Pruulmann-Vengerfeldt, Aljas; Runnel; Runnel et al.), we have analysed Estonian Internet users and found the six most common Internet user types (see fig. 1). Here one can see that, in general, uses can be divided into two categories – information related uses and entertainment related uses. Most active users can take advantage of the Internet and implement it for both kinds of uses, while most passive users use the Internet so little that their usage is not signified by either of these uses.

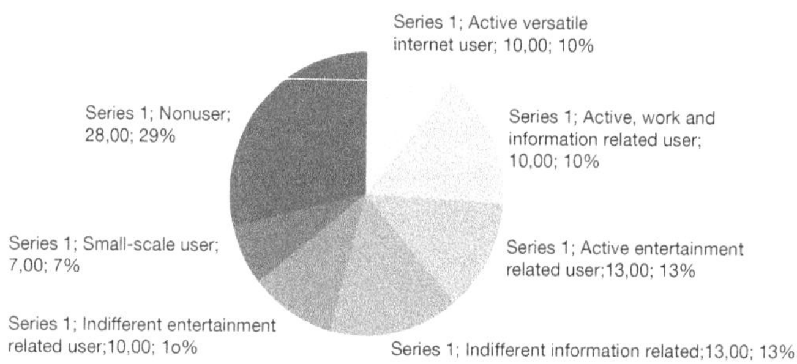

Figure 1: Estonian population in 2008, based on their relationship with the Internet.

Individuals tend to employ ICTs as tools with which to use existing entertainment services, to find information and to communicate. However, the use of the Internet to promote civic society and engaging in participation is marginal. The problem here is with the basic composition of user types. Active work and information-related users are most inclined towards democratic participation outside online environments. They use the Internet mainly to retrieve information and to perform work-related tasks. In addition, most of our interviewed heritage institution workers feel that they belong among the information-related users. This means that their conceptualisation of Internet services is also mainly centred on providing information. From the heritage institutions' perspective, they are the prospective users of various databases, probable readers of longer texts and are more likely also to ask for professional help and guidance.

At the same time, those who are more used to participating online, communicating and generating online content are among the younger users, and they can be mainly found among the entertainment-related user groups (see also Pruulmann-Vengerfeldt, Kalmus and Runnel). On the one hand, they are more likely to contribute, and to participate in competitions and various online activities. At the same time, previous research also indicates that due to their lack of literacy skills, they would prefer to have information in digested, interpreted and easily accessible ways. Information related to school work or other particular tasks is most expected (Laak). While more used to communication and entertainment, their critical skills are not that significant.

This indicates that the question of the roles of the consumer and the citizen also arises in the context of information and communication technologies. Although there are arguments that indicate inherent participatory potential in online technologies, which would give rise to active citizens, the actual uses of the Internet indicate a much more consumption-oriented behaviour. In different studies, consumers are traditionally associated with passive, mostly non-critical hedonism and the tendency to satisfy personal interests, whereas citizens are associated with active social thought and a sense of responsibility, enabling them to rise above narrow private interests (see also Gabriel et al.; Keller et al.).

The discussion above indicates that there are different types of expectations among Internet users. The playful and participatory potential is often coupled with less experience, less critical skills and more expectations towards easily digestible and accessible information titbits. At the same time, others expectations are more related to information retrieval, comprehensive databases and relevant knowledge made available for those who have skills to seek that information. To a certain extent, these demands are contradictory and pose a challenge to heritage institutions as to how to manage those expectations, especially given the circumstance of limited resources.

4. Heritage institution choices for user generated content

We have based the analysis on four organisations in Estonia, each of which has its own experience in participatory user engagement related to their different roles and functions as heritage institutions: first, the University of Tartu Library representing libraries, second, the National Archives of Estonia representing archives, third the Estonian National Museum representing the museum position, and fourth the Estonian Literary Museum representing a fusion heritage institution, since it serves the function of a museum, library, archive and research institute. Each of these institutions takes a different approach to their relationship with the public, and that approach is reflected also in their relationship to the digital environment.

4.1 Expert user with knowledge

It can be said that archives and libraries are more public service-oriented and that their attitude towards users resembles primarily a provider-client relationship – both provide a service for the public and act accordingly in their client relations.

In our analysis the first institution, the University of Tartu Library (UTL) has differentiated their users based on their 'relative proximity' to the institution. Academic staff from the University of Tartu are given the greatest possibilities to contribute – their opinion is the sole factor on which the library purchases access to online databases, and staff opinion matters when adding to the library's collections. Other users can also suggest books to the library, but their opinion has less weight. Ordinary users who are not institutionally affiliated to the library can leave anonymous comments on the library's forums where they are used, if possible.

The rating system of the library's books, implemented in the electronic catalogue of the library's ESTER[1] system, is available only to registered users, and the planned commenting option will only be accessible to those who have some sort of affiliation with the library. In addition, at this time the library does not have a clear vision as to what to do with the information potentially provided through the comments. The idea of collecting user comments seems to be considered mainly for the pleasure of the readers rather than the actual benefits it might have for the library collections.

The relationships with online users of the library's resources are complicated as university staff often uses online resources without the conscious understanding that these are provided by the library and their connections are to the University of Tartu (log-in information for database access is central with all university information systems) (Lepik). This has created a situation in which a significant proportion of library users have not, for a long time (or never), set foot in the actual building of the library.

> The main building of the library is seen as a studying and meeting place, but mostly useful for students. Some faculty members had lots of nostalgic memories about UTL (from times they were students of University of Tartu) but today they all have remained at a distance from the main building. (Lepik)

From the other side the amount of the materials in ESTER or other online resources provided by the University are huge, so with every search relevant information can be found in the databases. Employees of the University Library (Personal interview, February, 2010) used the fact that digital resources are not always very easy to find in order to strengthen their positions as experts. Only a correct search will give correct answers and results which presume users have

1 http://tartu.ester.ee/

studied information literacy courses beforehand, provided by the library to students and University staff. University Library employees argued that good literacy skills that can, in their view, only be obtained in connection to the library, are becoming crucial to survival in the information age.

4.2. Active users with good literacy and communication skills

The second institution, the National Archives of Estonia, is positively proud of the fact that access through their digital collection and demand-based digitisation has enabled them to host significantly more visitors to their online collections and thus opened the archival resources to the wider public. The deputy state archivist and director of the historical archives Indrek Kuuben states:

> We have opened a new room for researchers, the virtual research room. In the archive we have 20 workplaces for researchers; in a day 40-50 people go there.... But since we have opened the online research room, there are 60 people in the morning and perhaps 500 a day in the Saaga[2] genealogical database. So the use of archive materials has largely expanded. (I. Kuuben, personal interview, 8 April 2009)

While previously archives were hidden places, available for a limited number of researchers and writers who mediated the knowledge that they received from the archives to the general public, then with the advancement of the digital technologies, archives are using the opportunity to redefine their relationship with the public. There is an increasing interest in seeking personal, community or local area roots from the archives and digital technologies enable stronger personal relationships with the archival resources.

The databases are not easy to use and so in a way the user-generated content is created by genealogists and others interested in genealogy, who take digital archival materials and build on them additional layer of knowledge transcriptions and information networks. As Marlow et al. have also stated, people with good literacy skills have often started for personal organisational reasons and later moved on to the social benefits in order to help other users manage in the systems.

4.3. Communicative users with simple literacy skills

The Estonian National Museum and the Estonian Literary Museum look at the public not only as audiences of their exhibitions and consumer products and services, but also from the perspective of research disciplines, such as ethnology, folkloristics and anthropology. These disciplines have approached individuals and groups as subjects of research and as sources of collecting information.

2 www.ra.ee/saaga.

Online participatory options in the Estonian Literary Museum and the Estonian National Museum are more geared towards facilitating the dialogue with users – asking them to comment on and add to digital collections as a complex body of knowledge. As an example online database of Traditional Folk Calendar BERTA[3] asks people to provide their own ideas on how to celebrate traditional holidays. Until the spamming-robots conquered the initiative with their input, anyone could add, in a comment format, their knowledge and ideas concerning a particular national holiday. Here the comments interacted with each other and instead of remaining single contributions became integrated with the collection and the use of the database, making contribution easy for users with fewer literacy skills.

However, experience from the Estonian Literary Museum indicates that when participation is made too easy, this could also reduce the quality of the contributions.

> When contemporary school folklore was gathered, those contributions made anonymously and in the online environment were less thorough and well-written than those which were contributed in the class-room environment under the watchful eyes of the teacher. (Piret Voolaid, personal interview, 9 April 2009)

Another example from the Estonian Literary Museum illustrates the linking of the collections by users of the Kreutzwald's Century[4] online project, in which the user can explore history in a non-linear way, thus creating (though not leaving a record of) her or his own trail through literary history. Here, one potential application of user-generated content is to store the trails of the digital content users and provide them as potential pathways to those interested in the non-linear narration of literary history. A great future potential also lies in recommendation systems supporting participatory activities, which goes beyond the model of commercial providers (e.g. Amazon) of outlining similarities between products by providing a social recommendation system recommending relevant marked units from the collections or additions from the users, based upon the material in which a user has already expressed an interest. Such social recommendation systems may also provide recommendations based on both familiarity/similarity, also enabling the conscious comparison and connection of objects, stories or comments which perhaps represent different perspectives and use other people's recommendations to help in the connections.

Although the Estonian Literary Museum has provided many interesting and open online collections, the materials in the different databases are organised by archives, which inevitably makes usage complicated to many potential users as it needs their previous knowledge to make the right search in the right database.

---

3 http://www.folklore.ee/Berta.

4 http://kreutzwald.kirmus.ee.

The Estonian National Museum had a campaign that took place both online and offline, aiming to document everyday life in 2009 – "Give Museum a Day from Your Life". People were asked to document their April 14th, 2009, the 100th anniversary of the museum. Their contributions were included in the collections of the Estonian National Museum. Contributions to these kinds of initiatives need attention and time from the contributors. They know that the stories and pictures later became part of the museum's collections and that adds a sense of value and motivation to audiences to participate. At the same time, the topic remained simple enough as everyone can claim to be an expert on their own and their family's everyday lives. People could use the way of contributing that they found familiar, so there was 202 offline and 223 online contributions – in text, blog, video, photo or mobile positioning format. Before the public call for participation, museum staff performed exercises 'collecting' their own lives; these stories were provided as examples of different styles of participatory content to help people overcome the complications of starting and choosing the format.

The idea that editing existing content is easier than starting from scratch connects well with Mark Carnall's observation that online museums are typically very content-light, which makes it difficult to attach the contributions of the public to specific pre-existing structures. Contribution to online content is related to the network effect (Liebowitz et al.), meaning that the resource becomes more valuable when there are other people consuming the same good. Thus Internet users expect and like to contribute where others are and where some prior content exists. The more available information there is, and the more opportunities there are to link, add, comment on and tag the information heritage institutions have online, the more valuable the resource is for individual users.

The Estonian National Museum also ran a user-generated content experiment in the real exhibition space, where visitor participation was made easy. Visitors to the exhibition were given the opportunity to add free-form comments to the presented photographs using simple post-it notes and pens. Motivation for this experiment was provided through the promise of a prize draw in which participants could win a particular photograph as a printout for their personal use. Eighty percent of the comments contributed were expressions of emotions (such as 'beautiful', 'great', 'I like', and/or 'I would like to have that too').

However, from the heritage institution's point of view, more valuable were the remarks that indicated the new knowledge that people received from photographs, or where the exhibited photographs activated new interests or questions. In addition, a few corrections were made to the photograph captions that the museum had, for example: "It should be Artur Vasiksaat, because the name Vasikraat does not exist on Muhu". At the exhibition we also saw that through participation the visitor's role extended, as many analyses on exhibition participation have recently shown (Ciolfi et al.).

5. Discussion

Institutions try to apply dimensions of those relationships that are based on previous experience, and try to apply the same to users of the online environments. Hence the options for online materials in archives and libraries are geared more towards providing digital content, with the institutions expecting users to be engaged in the discussions about the provided materials separately from the digital collections. At the same time, the museums in our case study focused on collecting and adding the contributions of museum collection users, with more attention being paid on the dialogue and the added value of user discussions.

In recent years, heritage institutions have invested significant amounts of time and resources in the digitisation of their collections and providing open and free online access to them. Institutions are adjusted to the idea that physical and virtual collections are different, and that the latter can provide more users or visitors.

Massive digitisation and the opening of collections to online public access has become central to the daily work of heritage institutions, necessitating changes in organisation activities. At the same time the institutions use the same communication patterns both offline and online, which originate from the traditional role and habits of communication of the heritage institution, resulting in its perceived expert position. The latter is a barrier for the audiences in using online collections or participating in online activities. The main focus of the activities has been on making the collections available, not on analysing new needs or re-evaluating the activities in new situations. There is a growing competition between the heritage institutions for online visibility with their digital collections; often those who have thoroughly thought about what material is offered online and to whom it is offered find more success and gain a more stable user group.

There have been arguments (Carnall) that say that memory institutions have had significant obstacles that have stopped them from being online to a great extent. These obstacles have included the genuine fear that people would stop coming to museums if they could access museum collections online (ibid.). The examples of the National Archives of Estonia and the University of Tartu Library show this is well founded.

> The idea of becoming virtual might not be a pleasant one for some museums, especially not for art museums who cherish the ideal of the 'real thing' and its aura. But this development is inevitable because of the increasing digitisation of cultural heritage and the demand to make collections more accessible. Eventually, these trends will blur the differences between cultural heritage institutions, and in the long run these institutions will merge into one memory institution. A memory institution combines digital surrogates of the collections of archives, libraries and museums in rich interactive environments and allows access to the content regardless of the nature of the institution. The goal of the memory institution is to preserve this content for future generations and support its use and management over time. (Schweibenz)

These interlinked memory institutions, hailed by Werner Schweibenz above, show that important user motivation comes from the content itself. National museums and ethnographic collections particularly can claim to be 'living' museums where communities connected with the museum can add content to the collection, originating for example from the daily life of the community.

When looking at different Internet user types in Estonia we can see that most users consume information from the Internet. Only a small number of users participate in content creation. We have analysed ways in which heritage institutions communicate with the public and see the role of their audiences. It could be said that audiences are mainly handled as passive information consumers or users with good literacy skills. However, the offline activities of exhibition visitors show that people are willing to participate if the environment is familiar and the technology also supports their experiences of heritage and its interpretations.

Therefore, heritage institutions are facing challenges to find new audiences, in addition to enthusiast users, to activate online participation in producing and interpreting heritage, which also makes users knowledge producers and allows new personal interpretations of heritage.

Acknowledgements

The authors are grateful for the support of the Estonian Science Foundation, research grants Nos. 7162 and 8006, and target financed projects SF0180002SO7 and SF0180017SO7.

References

Sources

Focus group interviews. May 2009.

Personal interviews with employees of the Estonian National Museum, the Estonian Literary Museum and the National Archives of Estonia, April 2009, and with employees of University of Tartu Library, January-February 2010.

MeeMa. Survey "Me. World. The Media" carried out by the Institute of Journalism and Communication, University of Tartu. 2008.

Literature

Carnall, Mark. "Chasing the Online Audience." *University Museums and Collections Journal* 2 (2009): 37-39. Web. 28 Feb 2010. http://edoc.hu-berlin.de/umacj/2/carnall-mark-37/PDF/carnall.pdf.

Bruns, Axel. "Towards Produsage: Futures for User-Led Content Production." *Proceedings: Cultural Attitudes towards Communication and Technology (Tartu, Estonia, 28 June - 1 July 2006)*. Eds. Fay Sudweeks, Herbert Hrachovec and Charles Ess. Perth: Murdoch University, 2006: 275-284. Web. 28 Feb. 2010 http://snurb.info/files/12132812018_towards_produsage_0.pdf.

Carpentier, Nico. "A Short History of Participation in the Cultural Realm." *Digital Turn. User's Practices and Cultural Transformations*. Eds. Pille Runnel, Pille Pruulmann-Vengerfeldt, Piret Viires and Marin Laak. Peter Lang, 2012. 198-212. Print.

Cha, Meeyoung, Haewoon Kwak, Pablo Rodriguez, Yong-Yeol Ahn, and Sue Moon. "I Tube, You Tube, Everybody Tubes: Analyzing the World's Largest User Generated Content Video System." *Proceedings of the 7th ACM SIGCOMM Conference on Internet Measurement IMC'07. October 24-26, 2007. San Diego, California, USA*. 2007. Web. 28 Feb. 2010 http://citeseerx.ist.psu.edu/viewdoc/download?doi=10.1.1.119.1114&rep=rep1&type=pdf.

Ciolfi, Luigina, Liam J. Bannon, and Mikael Fernström. "Including Visitor Contributions in Cultural Heritage Installations: Designing for Participation." *Museum Management and Curatorship* 23.4 (2008): 353-365. Web. 28 Dec. 2010 http://dx.doi.org/10.1080/09647770802517399.

Davies, Sue M. "The Co-production of Temporary Museum Exhibitions." *Museum Management and Curatorship* 25.3 (2010): 305-321. Web. 1 Dec. 2010 http://dx.doi.org/10.1080/09647775.2010.498988.

Gabriel, Yiannis, and Tim Lang. *The Unmanageable Consumer*. London: Sage, 2006. Print.

Keller, Margit, Pille Pruulmann-Vengerfeldt, and Veronika Kalmus. "Quality of Life in a Consumer and Information Society: Defining Consumer and Information Society." *2008 Estonian Human Development Report*. 2009. 102-103. Web. 28 Feb. 2010 http://www.kogu.ee/public/EIA2008_eng.pdf.

Kuutma, Kristin. "Cultural Heritage: An Introduction to Entanglements of Knowledge, Politics and Property." *Journal of Ethnology and Folkloristics* 3.2 (2009): 5-12. Print.

Laak, Marin. "Heritage, User and the Environment: Rewriting the Narrative of the Literary Past in the Digital Space." *Digital Turn. User's Practices and Cultural Transformations*. Eds. Pille Runnel, Pille Pruulmann-Vengerfeldt, Piret Viires and Marin Laak. Peter Lang, 2012: 82-100. Print.

Lepik, Krista. "Tartu University Library as a Medium between Different Patrons." *Qualitative and Quantitative Methods in Libraries: Theory and Application. Proceedings of the International Conference on QQML2009, 26-29 May 2009, Chania, Crete, Greece*. Eds. Anthi Katsirikou and Christos H. Skiadas. 2009. Web. 28 Feb. 2010 http://www.isast.org/proceedingsQQML2009/PAPERS_PDF/Lepik-Interpretations_human_resource_management_in_digital_library_education_PAPER-QQML2009.pdf.

Liebowitz, Stan J., and Stephen E. Margolis. "Network Externalities (Effects)." *The New Palgrave's Dictionary of Economics and the Law*. MacMillan, 1998. Web. 28 Feb. 2010 http://www.utdallas.edu/~liebowit/palgrave/network.html.

Marlow, Cameron, Mor Naaman, danah boyd, and Marc Davis. "Position Paper, Tagging, Taxonomy, Flickr, Article, ToRead." *World Wide Web 2006 (WWW2006): Collaborative Web Tagging Workshop*. Edinburgh, Scotland, 2006. Web. 28 Feb. 2010 http://www.ibiblio.org/www_tagging/2006/29.pdf.

Nielsen, Jakob. "Participation Inequality: Encouraging More Users to Contribute." *Jakob Nielsen's Alertbox*. October 9 2006. Web. 28 Feb. 2010 http://www.useit.com/alertbox/participation_inequality.html.

O'Reilly, Tim. *What Is Web 2.0: Design Patterns and Business Models for the Next Generation of Software*. 30 Sep. 2005. Web. 28 Feb. 2010 http://oreilly.com/web2/archive/what-is-web-20.html.

Pruulmann-Vengerfeldt, Pille, and Agnes Aljas. "Digital Cultural Heritage – Challenging Museums, Archives and Users." *Journal of Ethnology and Folkloristics* 3.1 (2009): 109-127. Print.

Pruulmann-Vengerfeldt, Pille, Veronika Kalmus, and Pille Runnel. "Creating Content or Creating Hype: Practices of Online Content Creation and Consumption in Estonia." *Cyberpsychology: Journal of Psychosocial Research on Cyberspace* 2.1 (2008). Web. 28 Feb. 2010. http://www.cyberpsychology.eu/view.php?cisloclanku=2008060202&article=1.

Runnel, Pille. *Transformation of the Internet Usage Practices in Estonia. Dissertationes de mediis et communicationibus Universitatis Tartuensis 8.* Tartu: Tartu University Press, 2009. Print.

Runnel, Pille, Pille Pruulmann-Vengerfeldt, and Kristina Reinsalu. "The Estonian Tiger Leap from Post-Communism to the Information Society: From Policy to Practices." *Journal of Baltic Studies* 40.1 (2009): 29-51. Print.

Schweibenz, Werner. "The Development of Virtual Museums." *ICOM News* 3 (2004): 3. Web. 28 Feb. 2010. http://icom.museum/fileadmin/user_upload/pdf/ICOM_News/2004-3/ENG/p3_2004-3.pdf

Pille Pruulmann-Vengerfeldt, PhD, is an Associate Professor at the University of Tartu, Institute of Journalism and Communication and a Researcher in Estonian National Museum. Her Interests are internet user typologies, user-friendly online spaces as possible venues for participation and participatory applications for organisations. She is leading and participating in several national and international projects.

Pille Runnel, PhD, is a Research Director at the Estonian National Museum and Researcher at the project "Developing Museum communication in the 21st century Information Environment". Her research interests include museology, audience studies and Internet user studies, new media, democracy and participation. She also works as a curator and film festival director.

Agnes Aljas is a Research Secretary of the Estonian National Museum and PhD Student of the Institute of Journalism and Communication at the University of Tartu. She has MA from the University of Turku, where she studied ethnology and BA in ethnology from the University of Tartu. Her PhD focuses on the changes of the concept of culture.

New Media and Changes in the Forms of Cultural Transmission: The Estonian Experience

Marju Lauristin

Professor of Social Communication
Institute of Journalism and Communication
University of Tartu
Lossi 36, 51014 Tartu
Marju.lauristin@ut.ee

1. Introduction

At present, Estonia is witnessing the replacement of transitional culture by the late-modern cultural pattern of the modern network society. The reasons for such a cultural shift are affected by the trends in global culture, changes in the technological environment and the emergence of a new generation which grew up in a free and open society to be both the creators and consumers of culture. The present article explores how the rising importance of the new media can influence the forms of cultural participation and socialisation.

2. Characteristics of the media as mediators of culture

Researchers of communication have already for decades been addressing the question how the content and functions of culture are affected by the character of the media used for the creation and reception of cultural content. According to Marshall McLuhan the effect of the media on culture is of crucial importance because of the way in which people receive the message, what senses they use to receive it and how emotionally or rationally and how creatively they do it depends on the medium. Looking at the speed and scope of change in the forms of cultural transmission due to the expansion of the new digital media, McLuhan's famous sentence "the medium is the message" deserves new attention. To reiterate, McLuhan distinguished between the 'cold' and 'hot' media, depending on the level of creative input expected from the respondent. Thus, a book is a 'cold' medium, because the understanding of the text of the book, from seeing the abstract letter combinations to arriving at the undivided meanings encoded into them, requires active thinking from the readers and creation of their own mental images. TV, cinema and theatre are, in contrast, the 'hot' media because viewers receive prepared mental images, which do not require that much 'co-creation' or 'heating' of the imagination. The culture that has been built upon book reading presupposes a longer period of learning and a relatively deeper intellectual effort, as well as better creation skills and a greater ability to interpret abstractions than the audiovisual culture, which is more supported by emotional effects, whole pictures and images, and symbol-like behaviour that copies reality. Or, in other words, the written word-centred culture that uses the 'cold' medium is rational and favours education hierarchies, while the character of the reception of sounds, pictures and images spread via the 'hot' media is more emotional, open and democratic. What could be from this viewpoint the effects of the digital media on the content of cultural transmission? From one side, they demand high level of rationality and specific skills for reading texts and data, from the other – they offer opportunities to use a rich world of images and sounds.

Conventionally we could, based on the media that predominantly disseminated culture in a certain period, distinguish between four different eras in the development of Estonian contemporary culture: 'the era of printed word' (which began in the middle of the 19th century with the publication of the national epic *Kalevipoeg* and the first nationwide newspaper, initiated by J. V. Jannsen):[1] 'the era of radio' (radio broadcasting started in Estonia in 1926 and this new medium acquired nationwide popularity in the 1930s and prevailed as the news medium up to the 1970s); 'the era of TV' (Estonian TV was established in 1958, and since the 1960s was able to bring into the living rooms of Estonian mass audiences not only local but also an international world of audiovisual images due to direct access to Finnish TV channels). These traditional media are challenged in 'the era of the Internet', which started on the mass level in Estonia in the mid-1990s with the famous 'Tiger-leap' project. These developments in diversification of the Estonian media world are depicted on the following scheme (see Figure 1). The arrival of each of these new media has changed the content and ways of popular participation in culture.

1 The start of the national press era from the 2nd half of the 19th century was a crucial factor in the Estonian national awakening. It coincided with other important developments that facilitated direct participation of the broader public in modern national culture: the beginning of the nationwide song festival in 1869 and the establishment of the "Vanemuine" national theatre association.

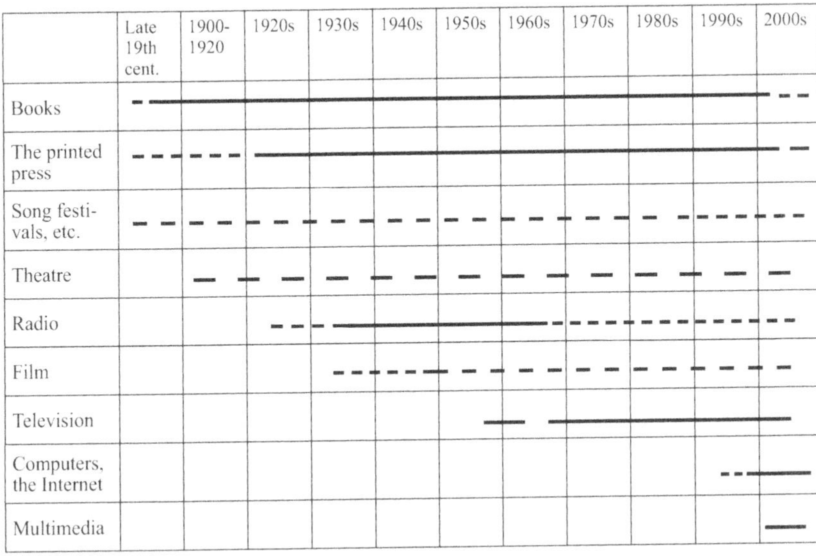

Figure 1. Prevalence of different media in disseminating culture in Estonia.

Following 'Mcluhanist' logic in understanding media culture, we can assume that the shifting of culture from one media environment to another not only initiates changes in the structure of the text but affects more deeply the meaning of cultural participation. When a message is recoded from one sign system into another (e.g. from a verbal system to an audiovisual one), the semantic fields shift, while at the same time the context of communication undergoes changes. Here, the effect of the medium is determined not only by its technical abilities to render a text in a certain coded format and using different senses for its reception (auditivity vs. visuality, linearity vs. hypertextuality, etc.), but also by the social qualities of the medium – the simultaneity or temporal difference in reception, individual vs. collective usage, institutionalisation of text-creation and reception, conventionality of certain ways of interpretation and usage and the canonisation of certain meanings, and the allowing of external control over the user.

Historically developed influential institutions are active in the cultural field, maintaining the stability of the normative meaning of culture, homogenising the expectations and judgements of cultural audiences, and preferring certain codes and ways of interpretation that prevail at the given time to other, more marginal ones. The effect of education as the means of canonising 'proper understanding' of culture was and continues to be the most important factor, but also art criticism as an institution has worked in the same direction (Bourdieu, *Cultural Produc-*

tion 161-175). In the era of audiovisual media, cultural space was overwhelmed by fashionable mechanisms (stars, cult films, cult music, etc.) and extremely simple and standardised cultural codes that often resembled folkloric mechanisms (myths, miracle fairytales). Mass culture, which came to triumph with the support of TV, radically changed the socialising mechanisms of the individual, trivialising meanings and values and transforming social situations (Meyrowitz 5-6). Already in the early 1950s, cultural critics indicated that the cultural relations of mass people, who are torn out of the book world that supports and shapes the uniqueness of personality, will then be shaped by craftily designed 'canned' emotions that make them defenceless toys in the hands of artificially evoked mass hysterias and euphorias (see Adorno and Horkheimer 41-42). Mass culture bore especially poisonous fruit in those cases, when it was mixed with the ideology of totalitarian societies and became a part of the state propaganda machine. In contemporary society, electronic mass media is one of the most powerful producers of the global cultural environment, filling people's senses with similar images, events, persons and stories.

The social mechanisms of the cultural field have been most thoroughly discussed in Pierre Bourdieu's treatment of culture as capital and cultural stratification as an important part and factor of social stratification (Bourdieu, *Distinction*). Book culture divided the cultural audience into firm categories according to their social status and education. In the television era, the opposition of elitist 'deep culture' and 'low' mass taste became even more prevalent (Gans 100-119). Opposing high culture to the low entertainment of the masses created a normative image of high culture with deep intellectual content as a rampart defending individual freedom against the dictate of commercialised as well as politicised mass culture. In Estonia and with other peoples who had been submitted to the Soviet regime, the individual valuation of culture was complemented by a collective mission: to maintain and protect national identity. Writers, poets, composers and producers were more than only cultural heroes – they became the leaders of intellectual resistance. The onslaught of market mechanisms on culture and media in the 1990s fractured cultural audiences, but also lowered the position of the creators of culture as the intellectual leaders of society. Artists and writers who had had to transform themselves into 'self-employed people' and popular actors who had to host entertainment shows and make TV commercials became the hostages of market relations. The main arena of commercialising culture was the audiovisual media with its serialised products and an abundance of commercials.

The digitisation of culture, the use of multimedia, the chance to surf global information flows accessible to everyone, and the spreading of social networks that cross institutional borders have, during the previous decade, much destroyed the classical mechanisms for spreading culture and dispelled mass consciousness.

Cultural stratification does not occur on the basis of traditional social and taste criteria, but is more and more related to the skill of using technological opportunities. Differing from the reception of books, newspapers and journals, but especially from TV and radio, the reception of new digital media is much less determined by traditions and normative interpretation schemes. Although at the beginning of the digital age much was said about the 'digital divide' as a new form of social stratification, today the overwhelming spread of the Internet has replaced this notion with a much more varied notion of 'digital literacy' (Runnel 29-31).

Studies of the effect of digital technology on the content of cultural processes have only begun (Laak 38-49). However, we can point out some principal differences between digital media and all previous channels for transmitting culture, which unavoidably entail principal shifts in the basic cultural mechanisms and social relations affected by culture (see Castells 327-375; Manovich 212-236; Liestøl, Morrisson and Rasmussen 298-321).

In the following table we have tried to highlight some hypothetical differences in the mechanisms of cultural transmission via traditional channels and new media:

Table 1. Differences in the traditional and digitised forms of cultural transmission.

Traditional cultural institutions and media (theatre, books, films, printed media, broadcasting)	**Interactive digital media, Internet networks ('new media')**
Important role of narrative as the form of transmitting culture	➢ Database-based forms of cultural transmission
Symbols have learned and fixed meanings, the way of interpreting cultural messages is specified by institutions, authorities and traditions of public text creation	➢ Broad possibilities for ad hoc interpretations of symbols and messages, freedom from authorities, traditions and norms
Fixed communicative roles, differentiating the communicator from the audience	➢ Merging of communicative roles ('produsers'), identity of the communicator and the audience
Social hierarchies and cultural barriers limit the accessibility of information and establishment of contacts	➢ Possibility to access any kind of information and contacts, disappearance of social and cultural barriers
Fixed identities and communities	➢ Changing identity, temporal virtual communities
Postponed reaction to messages, delayed effects and gratifications, high opportunities for reflections	➢ Immediate online reactions, ad-hoc effects and gratification, low opportunities for reflection

3. The increasing cultural role of the new media in Estonia

Beginning in the late 1990s, but especially after the turn of the century, the role of the Internet as a cultural channel rapidly gained importance. By 2010, Internet use had reached 100% among young people, more than 75% among middle-aged people and among retired people; it is on its way to rising to one third (see Figure 2).

	Autumn 2000	Nov 2001	Nov 2002	Nov 2003	Nov 2004	Nov 2005	Nov 2006	Nov 2007	Nov 2008	Autumn 2009	Autumn 2010
Whole population	28,2	32,5	43,5	46,6	47,4	52,8	59,4	65,5	66,6	72,5	73,5
Age											
15-19	73,7	82,7	89,3	87,3	88,2	91,7	97,1	98,3	98,8	100,0	100,0
20-29	43,7	50,8	69,5	77,2	76,2	81,2	89,4	94,0	96,9	96,0	97,4
30-39	30,9	40,6	57,7	62,5	58,6	65,8	77,6	87,8	87,4	91,4	90,0
40-49	22,8	23,3	35,0	38,7	45,8	55,4	57,6	69,8	69,4	77,9	75,8
50-59	14,6	15,4	25,2	27,2	29,7	32,9	38,1	43,1	45,3	59,7	60,7
60-74	3,6	3,7	6,7	7,5	6,2	8,9	13,5	14,5	15,9	23,3	30,1

Figure 2. Spread of Internet use by age group (share of users, %).

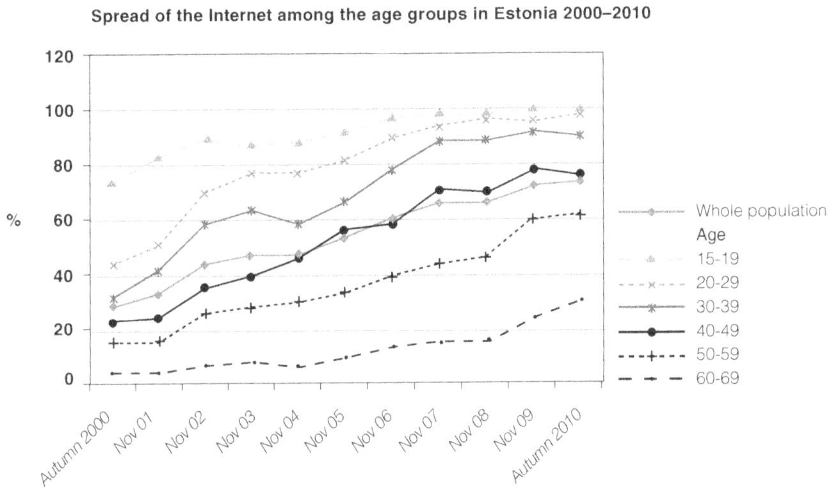

Source: *TNS Emor*

Together with the rise of the Internet as part of daily life an important cultural shift took place in post-communist Estonia – the effect of symbolic capital based on consumer values increased in interpersonal relations. A consumerist, brand-centred cultural model developed, characterised by taking people and things at their 'market value'. During the years 2002-2008, consumerism grew in Estonia at approximately the same speed as the spread of the use of digital media (see more in Kalmus, Keller and Kiisel).

The simultaneous spread of the Internet and consumerism significantly changed the context of social stratification. For young people particularly, in addition to material wealth, success in adopting new practices and technologies of communication and consumption became increasingly important as a factor of subjective social stratification (Lauristin, "Eesti ühiskonna kihistumine").

As in many other countries, as in Estonia, along with the rapid growth in Internet usage the importance of the role of books as a cultural channel has gradually decreased, although this change has occurred at a much slower tempo than the onslaught of the Internet (see Table 3).

Table 2. Decreasing interest in books among Estonians, 1979-2008.

	1979-1984	1987	1990	2002	2005	2008	2011
Reading of fiction: At least once a week (also: often, regularly)	57	50	48	28	30	22	24
Once or twice a month (sometimes)	25	29	32	33	30	34	34
A few times a year, still less frequently (very seldom, not at all)	18	21	19	39	40	44	42
Reads often professional (work or hobby-related) literature	24	?	26	31	31	21	22
Seldom, not at all		?	38	31	29	44	50
Home library							
Has no books at home, or has very few books (less than 50)	23	5,3	4,3	20	25	23	34
Has about a hundred or a couple of hundred books at home	26	40	31	37	41	40	40
Has several hundred books at home	35	27	27	34	28	29	19
Has about a thousand or more books at home	16	27	37	9	7	6	7
Buying books At least once a week (often, regularly)	62	74	76	5	7	7	3
A few times a year (sometimes)	29	24	22	38	39	32	30
Once a year or even less frequently (very seldom)	9	2	2	37	35	33	32
Practically does not buy books	-	-	-	19	18	28	35

Source: Estonian Social Sciences Data Archive, representative media surveys conducted by Institute of Journalism and Communication of the University of Tartu

When we compare the present situation to the 1980s and 1990s, when read literary fiction was among the favoured activities for all social strata in Estonia (Lauristin and Vihalemm), we can see that people who still enjoy reading fiction now form a relatively small group. For the majority the main cultural channels no longer include books but rather the TV and, increasingly, the Internet. The essential function of reading books has significantly changed as well. In the 1990s, more and more people preferred reading all kinds of non-fiction and self-help books to reading fiction. A particularly important qualitative change in the forms of cultural communication has taken place in the second half of the 2000s, when young people started predominately to use the Internet to read the news and for entertainment. The results of the Me. The World. The Media (MeeMa) survey in 2011 clearly show that not only young people but also middle-aged groups are rapidly moving from printed to online newspapers:

Table 3. Do you read printed or online newspapers?

	15-19	20-29	30-44	45-54	55-64
Mainly printed	14	16	27	42	53
Printed and online equally	40	45	43	40	31
Mainly online	32	28	22	11	7
I do not read newspapers at all	14	11	8	7	8

Source: MeeMa 2011[2]

Especially for the younger age group, the Internet is transforming into a multifunctional cultural environment, more and more often replacing traditional cultural channels.

Table 4. Have you stopped using traditional media channels for the following purposes?

	Age					
Has stopped...	15-19	20-29	30-44	45-54	55-64	Total
Reading printed newspapers	27	27	23	29	9	19
Watching films, programs on TV	18	16	8	4	3	8
Listening to music on the radio	24	24	10	9	6	12
Listening to news on the radio	24	25	15	12	7	15
Has not stopped any old media activities	54	55	65	72	84	69

Source: MeeMa 2011

2 MeeMa is the acronym for a series of nationwide surveys entitled "Me. The World. The Media", conducted by Institute of Journalism and Communication (University of Tartu) in 2002, 2005, 2008 and 2011. The surveys consist about 800 indicators reflecting lifestyle, values, social and cultural orientations, patterns of media usage, participation in economic and political life, wellbeing and consumer culture. Surveys use a combined method of written questionnaire and oral interview. The representative sample of the Estonian population includes 1000 Estonian language speakers and 500 Russophones between the ages of 15 and 74.

The spread of new media has, undoubtedly, already changed the cultural environment in several ways, especially the forms and possibilities of the cultural participation and self-expression of young people. In the on-going cultural communication in the digital environment, the difference between the 'traditional audience for culture' and the 'consumer of new media' is gradually losing its initial meaning. A sharp rise in the technological sophistication of cultural media dispels the border between elitist 'high culture' and 'low' mass culture (Lõhmus, Lauristin and Siirman). Through the social networks, the formerly exclusive club culture is accessible to a wider circle of people, while the sophisticated forms of computer art could not be confused with the ordinary products of the mass media.

Our data does not confirm the quite widespread opinion that the use of the Internet has a negative effect on cultural participation. Rather, an active Internet user is often also engaged in more traditional cultural activities (Figure 3).

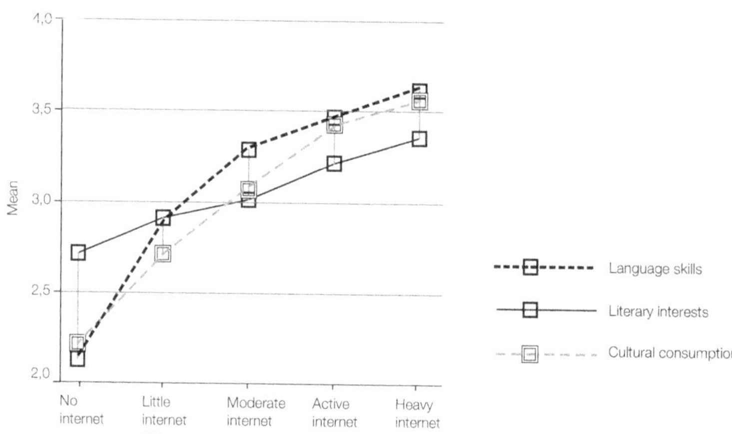

Figure 3. Active participation in culture increases with a rise in the active use of the Internet. Source: MeeMa 2011

However, one should ask whether and how the meaning of cultural relations and the subjective valuation of culture change, when the media, which had traditionally been used to communicate culture and participation in the cultural field, changed to digital.

4. The subjective meaning of cultural experience

As was stated above, changes in the media of cultural transmission entail changes in the meaning of cultural participation for the public. In the MeeMa survey of 2005 we attempted to clarify how people interpret the meaning of culture and its role in their lives. We offered our respondents 24 different statements concerning the notion of culture and asked in what way they agreed with each of them. Using factor analysis these 24 statements were reduced to six generalised cultural orientations (see Table 5). The first, most overwhelming orientation relates culture to personal development and education (new ideas, experiences, ethics). The second factor reflects the importance of culture to the essential aesthetic values of creativity, self-expression and harmony. The third factor contains statements that interpret culture as entertainment. The fourth factor indicates cultural self-conscience, competence, which means culture as a source of certain symbolic capital. The fifth factor highlights the role of culture in connecting people and shaping their (national) identity. The sixth factor refers to the cosmopolitan characteristics of taste and the importance of one's reference group, which supports expressions of personal taste.

Table 5. Six factors of cultural orientation (main component factor analysis, normalised Varimax rotation).

	1st factor Education	2nd factor Self-expression	3rd factor Entertainment	4th factor Competence	5th factor Identity	6th factor Distinction
Art offers new ideas	**0,675**	0,135	0,093	0,008	0,104	0,261
Culture shapes one's personality	**0,652**	0,287	0,151	0,305	0,188	0,053
Culture teaches ethics	**0,63**	0,298	0,031	0,173	0,312	-0,004
Art offers experiences	**0,544**	-0,015	0,416	0,006	0,309	0,152
Life is dull without art	**0,428**	0,337	0,295	0,193	0,257	0,191
The essence and harmony of life are expressed in art	0,16	**0,74**	0,059	0,094	0,049	0,064
The inner self is expressed in art	0,036	**0,63**	-0,061	0,086	0,172	0,434

	1st factor Education	2nd factor Self-expression	3rd factor Entertain-ment	4th factor Competence	5th factor Identity	6th factor Distinction
Freedom of the creative person is important in art	0,255	**0,629**	0,241	0,069	0,091	0,072
Art helps us understand life	0,406	**0,524**	0,16	0,362	0,117	-0,012
The value of art depends on the person's mood	0,143	0,111	**0,775**	0,091	0,081	0,044
Culture is entertain-ment	0,011	0,082	**0,705**	0,062	0,225	0,214
I don't like arguing about culture	0,308	0,105	**0,508**	0,261	0,287	0,101
Art is the best enter-tainment	0,087	0,178	**0,438**	0,438	0,295	-0,066
I follow reviews and critical articles	0,036	0,059	0,053	**0,781**	0,153	0,143
I want to keep up with the times	0,2	0,132	0,181	**0,7**	0,103	0,091
I can value art	0,036	0,386	-0,064	**0,448**	0,086	0,385
I'm interested in discovering other cultures	0,3	0,38	0,125	**0,422**	0,055	0,128
Interest in culture unites people	0,324	0,059	0,067	0,223	**0,627**	0,161
I enjoy the atmosphere of perfor-mances, etc.	0,212	0,007	0,298	-0,026	**0,627**	0,181
Culture is the basis of national identity	0,246	0,218	0,191	0,198	**0,602**	0,049

	1st factor Education	2nd factor Self-expression	3rd factor Entertain-ment	4th factor Competence	5th factor Identity	6th factor Distinction
I feel close to traditional culture	-0,053	0,447	0,12	0,264	**0,524**	-0,119
My friends appreciate my taste	0,053	0,223	0,074	0,189	0,254	**0,686**
I'm more interested in translations than in Estonian literature	0,326	-0,007	0,313	0,079	0,01	**0,54**
I like to read books or watch films in their original languages	0,236	0,066	0,414	0,078	-0,059	**0,496**

Source: *MeeMa 2005*

Assuming that the composition of these factors indicates important distinctions in the personal meaning of cultural participation, we can ask how these specific cultural orientations are related to patterns in the use of media by different social categories. Our specific interest here lies in the effects of the traditional vs. new media.

5. Connecting cultural relations with the use of new media

With the ever-increasing multifunctionality of the Internet for the young audience, it becomes crucial to know how the cultural orientations of active Internet users differ compared to those people who prefer to use traditional cultural channels. Analysis based on the results of our survey from 2005 allow us to cast some light on the differences in cultural orientations between people who actively practice culture, e.g. who often go to concerts and theatre and read books, with those people who use the Internet to download music and films and play fantasy and role-play games. We can see (Figure 4) that participation in culture via computers is mainly connected with entertainment, and that heavy Internet users clearly have different tastes compared to the people who prefer traditional cultural media. At the same time, people who go in for hobbies involving self-expression (photography, films and videos, playing musical instruments) are again different, although their differences are more related with identity and self-expression than entertainment. Reading books is still most closely related with identity, and competence prevails in the cultural relations of people who visit theatres and concerts.

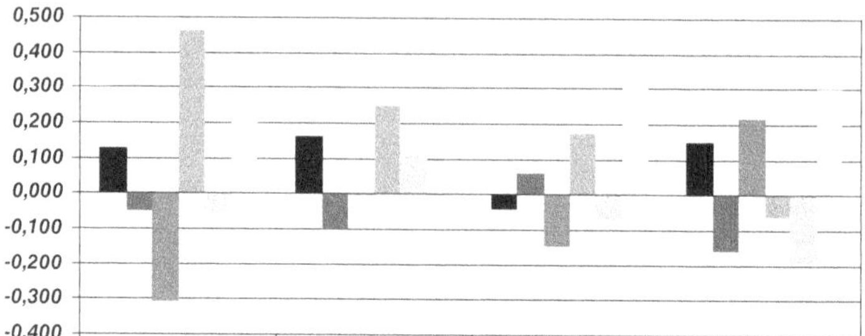

	Participation in cultural publics	Reading	Traditional cultural hobbies	Computerized hobbies (games, blogging etc)
F1 Education	0,126	0,164	-0,042	0,151
F2 Expression	-0,049	-0,099	0,060	-0,158
F3 Entertainment	-0,307	-0,002	-0,147	0,217
F4 Competence	0,461	0,251	0,175	-0,063
F5 Identitity	-0,064	0,117	-0,076	-0,195
F6 Distinction	0,218	-0,006	0,372	0,299

Figure 4. *Different meanings of culture for the groups who are actively involved in traditional and computerized forms of cultural participation (mean scores of the factors represented in Table 5).*

Members of traditional cultural publics and practitioners of traditional cultural hobbies are characterised by greater reflexivity and higher awareness of the cultural realm (a higher factor of competence). Those characterised by a high level of computer-related cultural practices are keen to show their distinction from 'mainstream' cultural orientations, and like to emphasise their special preferences and choices. At the same time users of computer-based cultural forms have stronger orientation to entertainment. For example, the responses of people who regularly listen to music on their computer, compared with active theatre- and concert-goers, highlight the entertaining and exciting side of culture, whereas for people who listened to live performances of classical and pop music, the meaning of cultural participation is more related to identity, self-expression and deeper competence. However, as these active computer users are, mainly, young people, our data does not allow us to say whether their attitudes can be explained by their age or by the impact of new media.

6. Concluding remarks

The rapidly spreading use of digitised media in the 2000s means much more than only a technical novelty. Although much has been said about the shift from the printed word to digital culture, the scope of this cultural change is still far from being understood. Compared with the printed word, entirely different principles of rendering and valuing cultural content are valid in the digital environment. Many self-evident meanings and relationships disappear and participation in the global virtual environment intensifies cosmopolitanism and sub-culturalisation. Proceeding from the above, we could deduce a number of problems with the preservation of institutions, practices and expectations that have developed in the previously traditionally book-centred culture of Estonia bearing in mind that communication in the digital format will probably become prevalent in all spheres of life. Some of these changes can already been seen in youth culture, but the majority of them are growing and are at present in a hidden, embryonic state. One of the first troubling issues was related to the question of whether the online press is meeting our traditional expectations of the press. A similar process is starting with books, and the i-revolution will be forthcoming at school.

However, there is no reason to fear that new media will entirely push aside the traditional cultural forms and media. We can already see that the essential patterns of cultural relations, the motives and forms of cultural participation are relatively conservative. The spread of the Internet has not dislodged traditional forms of culture; on the contrary, the more enthusiastic participants in traditional cultural forms also stand out as having better skills and more active use of digital media. The mobile phone does not replace concert halls, and the computer does not replace choral singing or folk dancing. Rather, the functional differentiation between different media, as well as new relations among them, are growing and multimediality will increase in traditional forms of culture. A question of survival for culture will be whether the young generation will find their way, using the virtual environment, to the older strata and essential values of their own culture, which are at present accessible mainly in books, theatre and concert halls, and art exhibitions.

Estonia has been one of the most enthusiastic and daring among the post-communist transitional states in using the possibilities of new media. As both the deeper positive and negative effects of this process are still in the latent form, it is too early to talk about the accompanying cultural changes as something that has already been proven. Initiated in the 2000s, the rapid shift to the use of new media not only as the main channel of communication and information, but also as an environment for cultural participation will continue in the coming years. It will involve, in addition to young people, more and more people of all ages. The triumph of new media will affect changes in cultural institutions and cultural participation practices and their content in a similar way to that which occurred in the 2000s in banking and many other fields of economic and political activity. While up to now, the shift to digital culture has proceeded relatively slowly and in the form of repeating the known cultural forms on the platform of new media, tomorrow we may face much larger global changes that affect the existential problems of a small national culture's survival in a much deeper and more unexpected way than before.

Acknowledgements

This article was supported by the Estonian Science Foundation grant GHZU no. 8329 "Changing Cultural Dispositions of Estonians through the Four Decades: from the 1970s to the Present Time" and the target-financed research theme SF0180017s07 "Estonia as an Emerging Information and Consumer Society: Social Sustainability and Quality of Life".

References

Adorno, Theodor, and Max Horkheimer. "The Culture Industry: Enlightenment as Mass Deception." *The Cultural Studies Reader.* Ed. Simon During. London: Routledge, 1993 (1944). 29-43. Print.

Bourdieu, Pierre. *The Field of Cultural Production*. London: Routledge, 1993. Print.

Bourdieu, Pierre. *Distinction.* London: Routledge, 1984. Print.

Castells, Manuel. "The Culture of Real Virtuality." *The Rise of the Network Society.* Blackwell, 1996. 327-375. Print.

Gans, Herbert J. *Popular Culture and High Culture: An Analysis and Evaluation of Taste.* New York, NY: Basic Books, 1999. Print.

Kalmus, Veronika, Margit Keller, and Maie Kiisel. "Emerging Consumer Types in a Transition Culture: Consumption Patterns of Generational and Ethnic Groups in Estonia." *Journal of Baltic Studies* 40.1 (2009): 53-74. Print.

Laak, Marin. *Kirjandusajaloo mittelineaarsed mudelid. Teksti ja konteksti probleeme digitaalses keskkonnas. Dissertationes litterarum et contemplationis comparativae Universitatis Tartuensis* 4. Doctoral dissertation Thesis. University of Tartu, 2006. Tartu: Tartu Ülikooli Kirjastus, 2006. Web. 19 Mar. 2012. http://dspace. utlib.ee/dspace/handle/10062/632.

Lauristin, Marju. "Eesti ühiskonna kihistumine." *Eesti elavik 21. sajandi algul: ülevaade uurimuse* Mina. Maailm. Meedia *tulemustest.* Eds. Veronika Kalmus, Marju Lauristin and Pille Pruulmann-Vengerfeldt. Tartu: Tartu Ülikooli Kirjastus, 2004. 251-285. Print.

Lauristin, Marju, and Peeter Vihalemm. "Raamatunõudlus ja kirjanduskultuur." *Keel ja Kirjandus* 7 (1986): 385-393; 8 (1986): 459-486. Print.

Liestøl, Gunnar, Morrison, Andrew and Rasmussen, Terje, eds. "Digital Media Revisited: Theoretical and Conceptual Innovations in Digital Domains". *MIT,* 2004. 298-321. Print.

Lõhmus, Maarja, Marju Lauristin, and Eneli Siirman 2009. "The Patterns of Cultural Attitudes and Preferences in Estonia." *Journal of Baltic Studies* 40.1 (2009): 75-94. Print.

Manovich, Lev. *The Language of New Media.* Cambridge, MA: MIT Press, 2001. Print.

McLuhan, Marshall. *Understanding Media: The Extensions of Man.* London: Routledge & Kegan Paul, 1964. Print.

Meyrowitz, Joshua. *No Sense of Place. The Impact of Electronic Media on Social Behavior.* New York; Oxford: Oxford University Press, 1985. Print.

Runnel, Pille. *The Transformation of the Internet Usage Practices in Estonia. Dissertationes de mediis et communicationibus Universitatis Tartuensis* 8. Tartu: Tartu University Press, 2009. Web. 1 July 2012. http://dspace.utlib.ee/dspace/handle/10062/14292.

Marju Lauristin, PhD, is Professor of Social Communication at the Institute of Journalism and Communication in University of Tartu, Estonia. She is President of Estonian Sociological Association and has been 2008-2011 the editor of the Estonian Human Development Report. Books: *Return to the Western World: Cultural and Political Perspectives on Estonian Post-Communist Transition,* 1998; *The Challenge of the Russian Minority,* 2002; *Estonia's Transition to the EU,* 2010.

Reaching and Including Digital Visitors: Swedish Museums and Social Demand

Tobias Olsson

Anders Svensson

School of Education and Communication
Jönköping University
Box 1026
551 11 Jönköping
Sweden
tobias.olsson@hlk.hj.se
anders.svensson@hlk.hj.se

1. Swedish museums and digitalisation

How can museums make use of digital media? What strategy could they apply in order to reach out to and involve the public using the help of the Internet's specific qualities? These questions make up central themes for this chapter, as it is our desire to present a two-step strategy for museums to apply in their efforts to involve the public. The approach has been developed as part of a Swedish research and development project, and as a consequence, the chapter starts with a short review of the Swedish policy context as it relates to the digitalisation of museums. From this point of departure we make a detour through relevant and practically useful theories before developing our two-step, hands-on model of how to go about digital development.

The contemporary digital media environment offers new opportunities to all kinds of societal institutions, not least to museums. A short list of a couple of digital features that might be of great value to museums would include their ability to help preserve, archive and distribute information (Aljas and Pruulmann-Vengerfeldt). Such a list would furthermore include a number of new presentational opportunities brought about by new media's interactive features (Buskquist; Jensen). It would certainly also include the ways in which their network-like logic (Barney) can help out in creating a network of museum networks that jointly presents their collections, which is a real case in the efforts to create *Europeana* (http://www.europeana.eu/portal/).

Looked upon from this point of departure, it is unsurprising that digital media have become included in many debates concerning how institutions dealing with cultural heritage in general, and museums in particular, can make use of the them. This debate has been played out at research conferences (for instance the annual Museums and the Web conference) and in academic publications (cf. Cameron and Kenderdine; Keene; Lang et al.), although it has also popped up in public debates, and not least within cultural policy.

When it comes to cultural policy concerning museums and digitalisation, Sweden is a case in point. During spring 2009 the results of the government's *"Kraftsamling! – museisamverkan ger resultat"* [Mobilisation – cooperation between museums makes results] (SOU) investigation were presented. One striking result of the investigation was that it managed to make use of as many as 40 (!) different constructions including the word "digital" in its 180 pages. A quick analysis of these constructions reveals that the word "digital" was added to a number of processes. They included anything from "administrative coordination" between museums, archives and libraries, to digitalisation of the museums' collections and museums' potential interactions with users through digital media.

In addition, when the Swedish government presented its culture bill, in September 2009, these "digital ideas" were clearly evident (Regeringens proposition). The word "digital" was mentioned 77 times in 158 pages, including attachments, and the bill made various connections between "digital" and "museums". For instance, the bill stated that: "Digital development offers [...] new opportunities for citizens and researchers to access cultural heritage" (Regeringens proposition 47, authors' translation from Swedish) and specifically noted the important fact that "digital technology makes it possible for the user to become co-creative" (ibid., authors' translation from Swedish). In other parts of the bill the Swedish government pointed to opportunities for institutional coordination offered to Swedish museums by digital technology (Regeringens proposition 46 ff.), but also noted such possibilities at the European level through *Europeana* (Regeringens proposition 47, 146).

Despite the obvious attention paid to digital media within Swedish cultural policy, the reality is that Swedish museums' appropriation of digital technologies has so far been less successful. This can be sensed in the abovementioned policy texts (cf. Regeringens proposition 49), and was also made explicit by an anonymous insider within the area during an interview. His view of the digitalisation of Swedish museums was quite obviously very negative as he stated: "Since 1995 Swedish museums have spent approximately one billion Swedish Crowns on digitalisation, but so far the museums can't communicate with one another, and to an even lesser extent with the general public."

The government bill, however, also has the aim of mobilising within the field. In 2010 the Swedish National Heritage Board was given responsibility for administering and developing *K-samsök*, a system for coordination of Swedish museums' digital information. *K-samsök* is not a website, nor a portal, but rather a "technical model that can distribute digital information between museums and to interested people and institutions" (Regeringens proposition 49, authors' translation from Swedish). Based on this "technical model" Swedish museums have started to try to link information from various museums into the same search system (www.raa.se).

2. A regional initiative – MINT

In this context, as the authors are about to contribute to a regional research and development project (see below), there are two important points on which to focus. Firstly, despite the attention paid to coordination of the museums' digital efforts, the digitalisation of Swedish museums is not a centralised project. *K-samsök* is rather an effort to harvest what museums make digitally available, although without prescribing any specific information standard. This also means that *K-samsök* does not prescribe the way in which local museums should go about the process of digitalising.

Secondly, the policy and general debates on digitalisation of Swedish museums have mainly treated it as an issue relating to museums as institutions (concerning coordination within and between museums), or as an issue that concerns museums' efforts to make their collections available online. As a consequence, very little interest has been paid to museum publics: the students, researchers, pupils or general public to whom the museums' digital material is made available. So far, this public actually appears as a phantom within the debates – an invisible, anonymous crowd that is supposed to make use of the digital information offered by museums. The debate has had very little input from the public in terms of what they want from museums, and to an even lesser extent information about how members of the public might want to participate as users of the digital environments offered by museums.

Within this context, the MINT project (*Museer, IT och Internationalisering i Jönköpings län* – Museums, IT and Internationalisation in Jönköping county) can be understood as a regional effort to digitally develop local museums. The project deals with the various dimensions of digitalisation. It is an effort to coordinate regional museums' activities and to make the museums' collections available online, which is very similar to the efforts made by Swedish museums in general. However, what is new with the project is the fact it is also concerned with developing the museums' interaction with the very often forgotten users. More specifically, it is a research and development project that aims to involve users in the development of museum public interfaces, websites and/or portals. In order to do this, the project makes use of two separate but interrelated techniques for user inclusion.

The approach firstly aims to be sensitive towards the web's possibilities for user inclusion. The main strategy here is to analyse best web practice – i.e. websites that have been successful in including users – brought from various societal spheres: commercial, civic, public, etc. An equally important part of the approach is, secondly, to both acknowledge and to account for users' deliberate and reflexive preferences when it comes to museum digital applications. With an analytical eye on best practice, a wide range of actual and potential users will be interviewed in focus groups aimed at grasping and analysing social demand (see below) in terms of museum digital applications.

In order to explain the approach this paper will discuss the project's theoretical foundation and view of museum digital applications. Thereafter the project's two separate, but interrelated, empirical strands will be described and discussed.

3. Museums as parts of the public sphere

Museums are usually described in various ways within public debates. Sometimes they are described as preservers of cultural heritage, in other contexts they are understood as educational institutions, and within still other contexts they are mainly looked upon as exhibition halls. These different views of museums are both frequent and relevant. Less frequent, but equally relevant, are descriptions of museums as parts of the public sphere, which is the theoretical frame of reference for the approach developed in this project.

In his book *Theories of the Information Society* (161-201) Frank Webster analyses the public sphere within the emerging information society. Using Jürgen Habermas' theory of the public sphere (Habermas) as a theoretical point of departure, Webster discusses connections between the idea of a public sphere and public service media. In Webster's analysis there is as clear connection between a media sphere governed by the principles of public service (a non-commercial space for information and public debates) and the idea of a public sphere characterised by civic inclusion and free access to information and debates. Hence, in his description of the institutions that are fundamental for a functioning public sphere, public service media – such as TV and radio – are unsurprisingly understood as essential. What is more interesting in his analysis in this case, however, is the fact that his list of the institutions that are considered valuable for a functioning public sphere also includes libraries, government information services and – above all – museums. Apart from the fact that museums traditionally have had public funding and low entrance fees for all citizens (182 ff), Webster identifies additional characteristics in museums as institutions that make them qualify for the public service sector:

> [A]n ethos of public service pervades museums [...] with curators and other staff upholding a professional commitment to provide and protect the collections in the interest of the general public (183).

Considering the title of Frank Webster's book – *Theories of the Information Society* – it is somewhat surprising that his analysis does not expand to include reflections on the significance of digitalisation for the museum's role in the public sphere. This would have been a very relevant endeavour considering museums' new, digital interfaces to the general public. Fortunately, however, other researchers have touched upon this issue. The Norwegian media researcher Hallvard Moe, for instance, has recently connected the idea of a public sphere to public services in the era of digitalisation. Within his analysis he also includes museum online services as parts of contemporary public services (Moe 330). Another example can be found in media scholar Graham Murdock's argumentation for a digital commons. Murdock argues for the need to build a digital commons to counterbalance commercial efforts to turn digital media in general and the Internet in particular into spheres for commerce and consumption. For Murdock a strong and vital public service sector is an important antidote to such a development and one part of that antidote consists of digitalised museums (16).

4. The museum's digital users as citizens

There are obviously good reasons to understand museums as parts of the public sphere. There might even be additional reasons to do so today when considering that digitalisation expands their interfaces to the general public. Such a positioning means that users should be understood as citizens rather than as consumers. There are of course many different ways of conceptualising citizenship (cf. Beiner; Isin and Turner), but in this context it suffices to say that a citizen is a person with civil, political, social (Marshall) and cultural (Stevenson) rights and obligations in society. During the last decade social science has also started to pay additional attention to the citizens' right to participate in both face-to-face and mediated public spheres (cf. Dahlgren for a recent overview), which is of specific relevance to museum digital interfaces.

One important consequence of positioning the user as a citizen has been made obvious in Canadian media researcher Mark Raboy's (Raboy and Abramson; Raboy et al.) analyses of social demand. Mark Raboy makes a clear distinction between consumer (or market) demand and social demand. He relates the latter concept to the public sphere and to issues concerning citizenship. According to Raboy, public service institutions should function according to the principal of social demand. Looked upon from this angle, their major obligation is to be there for the citizens. Commercial institutions, on the other hand, operate in the market. Their major obligations are to, first, survive as companies and, second, to bring revenue to the owners. This places specific constraints on their operations, as their activities need to be commercially viable, which in essence means that their efforts must try to meet the consumer's immediate demands. Mark Raboy argues (Raboy and Abramson; Raboy et al.) that the different institutions' various logics – the logic of public service and the commercial logic – ought to impose different premises for evaluation.

Another important part of Mark Raboy's argumentation is that the two kinds of institution, despite the differences between them, have usually been evaluated according to the same premises. Most often "the success" of commercial institutions, as well as public service institutions, has been evaluated by the use of quantitative measures. Following Ien Ang's famous criticism (Ang) of how market driven audience measurement makes media audiences sellable goods, Mark Raboy finds quantitative audience measurement problematic in general, and specifically problematic for evaluating public services. These statistical constructions of audiences make consumer demand out of what ought to be citizens' social demand:

> The market demand model takes form through the extensive and constantly evolving apparatus set up to gather audience ratings and measurements in order to generate the manageable figures needed to sell audiences to advertisers, determine programme popularity, and other aspects of the business of electronic media. The model is based on gauging the quantity and demographics of persons consuming given programming; the notion of public interest it encodes is very much based on an atomized conception of the viewing audience or community of users as individuals who come into being when the medium is in use [...] and whose interests are therefore served almost exclusively by finding programming which they do consume, for whatever reason. Therefore, while cultural policy-makers attempt to take into account citizenship considerations and non-market models of the public interest – of something other than "what the public is interested in", that is, commercial success – existing data sets are insufficient and lead them to fall back on market demand in conceptualizing the public in whose name they regulate. (Raboy and Abramson 335-336)

The extract specifically addresses media audience measurement, but the criticism is equally valid for other public service areas, such as museums. Against this backdrop of criticism, Mark Raboy outlines alternative approaches to the study and evaluation of public service media, approaches that are more careful to acknowledge the specific requirements involved in offering services to citizens rather than consumers. The approach aims to grasp social rather than consumer demand. It is critical of audience measurement and suggests instead qualitative methodologies that aim to grasp citizens' often-subtle preferences in terms of media.

In addition, museums traditionally apply these problematic audience measurements for evaluation. In its simplest and most evident form, this is what is done as museums measure success in terms of the number of visitors. These measurements are almost constantly applied by all kinds museums, and in some cases they even publish visitor numbers as a measure of success: "We have more than 300,000 visitors a year". This logic also applies to the museum's digital versions. From insiders we have for instance learnt how the web traffic to museums is used as a measurement of success, and furthermore that movements within these websites are used as indications of what digital services users want from museums. These kinds of data are of course valid and useful in some respects. However, if we are seriously interested in understanding museums as institutions within the public sphere, institutions that answer to citizens and public interest, these data of popularity are insufficient for evaluating their activities.

5. Social demand for the real – best practice and user involvement

The notion of social demand offers a useful framework for our research and development project, which aims to involve potential users in the development of museums' digital interfaces to the general public. The project is built on two different but interrelated foundations. On the one hand a study of best practice in the area of user involvement; on the other, a study of actual and potential online museum users. The latter study aims to take citizens' points of view and preferences into account when developing digital interfaces for regional museums.

5.1 Best practice: Principles and illustrations

According to the *Business Dictionary* (www.businessdictionary.com), best practice is defined as "(m)ethods and techniques that have consistently shown results superior than those achieved with other means, and which are used as benchmarks to strive for". The concept aims to develop local and personal entrepreneurial experience into universal knowledge, benchmarks and best practice, i.e. the concept aims to make it possible for companies to replicate other companies' successful practices. Despite its business and corporate origin, the concept has been used in such differing sectors as research methodology (Costello and Osborne); medicine (Grol and Grimshaw); learning communities (DuFour and Eaker); and schools (Codrington).

Several aspects of the concept have been criticised. Smith and Sutton object that the term connotes an authoritative finality implying a sense of stasis. Real practices evolve and improve over time. In addition, the concept "best" has universalising claims, implying that what works well in one organisation is also likely to work well in others. Purcell stresses the importance of understanding the circumstances in which practices are applied. Dana and Smyrnios notice the lack of empirical evidence for what practices that in fact are the best of all practices are implemented at different occasions. Their suggestion is to continuously employ expressions, insights, principles or lessons, alternative promises or proactive practices until verifiable evidence is obtained that warrants the description of practice as "best practice".

Despite the problematic transformation between institutions and contexts, and the difficulties in deciding whether a practice is best practice or not, we will still use the term "best", although our general understanding of the concept tends to be *better* or just *inspiring* practice. Since the objects of interest for us are websites, we will use the concept *best web practice*. By this we mean, firstly, websites brought from various societal spheres: commercial, civic, public, etc.; and secondly, websites that have been successful in including users. Our example here is drawn from the commercial sphere: the Swedish web company Moderskeppet (www.moderskeppet.se).

5.1.1 Moderskeppet – a corporate best web practice for museums to covet

The material on which this best web practice analysis is based are data gained from interviews with Moderskeppet's founder and CEO, a design analysis of the website, and textual analyses of Moderskeppet's blog. The best practice analysis focuses on four of Moderskeppet's practices: (1) the company's conception and knowledge of their visitors and users, (2) the gauging of their competitive strengths, (3) the design of the website, and (4) the appropriation and utilisation of different digital communication applications, i.e. social media. These four practices have been crucial in making the website popular among users to the extent that they – on their own initiative – market the website and the company within their personal, both physical and digital, social networks.

In the following, we will describe each practice. We will also suggest how these practices could be transformed from the corporate to the cultural heritage context. It is our opinion that museums can both use Moderskeppet's practices, after having adopted them to local circumstances, and benefit from them as best web practice in their work involving the audience via the web.

5.1.2 Commercial vs. public

As well as being a commercial enterprise, which distinguishes Moderskeppet from public museums, the company differs from museums in general by being almost completely dependent on the web. Armstrong and Kotler refer to such companies as click-only-companies. Museums, on the other hand, are generally first and foremost physical institutions for whom digitalisation means creating a complementary activity for display on the web. In Armstrong and Kotler's terms museums can consequently be compared to click-and-mortar-companies. In Moderskeppet's case the business is what is communicated on the web. For museums the web is often used to publish a digital booklet presenting its physical operations.

As a click-only-company Moderskeppet is forced to improve the website to user adapted and communicative perfection according to the devise: the web or life. To avoid a further costly and ineffective digitalisation, museums with visions reaching beyond the we-have-to-be-on-the-train level should have the same irrevocable and almost desperate approach. Despite this, some vital basic questions must be asked, for instance: What needs among our primary and ordinary visitors are satisfied by our presence on the web? And also: what type of less frequent visitor could possibly be reached through the web?

5.1.3 Getting to know users

As a newly established company, Moderskeppet took it for granted for some time that their visitors and users were identical to its staff, i.e. first and foremost men in their 30's with similar preferences. The communicative logic became, "this is what *we* want and consequently have to communicate". However, it is fairly obvious that communication intended for target group X becomes ineffective if most of the receivers are characterised by Y and Z. The crucial moment at which the website was improved to standards of good web practice was when Moderskeppet decided to examine the composition of visitors and users. This study showed that the users were not only quite equally distributed in terms of gender, but also of far more widely differing ages and educational backgrounds than imagined. These facts had immediate implications, firstly for the content and framing of communication and, secondly, for the staff and both the composition and division of tasks. Meeting visitors and users through the website, by performing in front of the camera or writing personal blog postings, proved to establish personal relationships between different members of staff and certain groups among the users. Consequently, the plurality of users demanded plurality among staff, something Moderskeppet themselves call the "Backstreet Boys effect".

5.1.4 Making use of competitive strengths

Moderskeppet's business is education. They produce instructive texts and video-based lessons on digital photography and editing. One of the most crucial proficiencies within education is to be pedagogical, i.e. to enable people to improve their understanding and management of subjects. Moderskeppet's employees are all chosen for their pedagogical qualities and their passion for teaching. Pedagogy is the first of Moderskeppet's competitive strengths. In order to make new visitors discover the pedagogical qualities of the educational material, almost everything on the website is gratuitous, i.e. free of charge. Generosity is the company's second competitive strength. Moderskeppet's competitors have tried to match these two strengths. They have copied the idea of offering educational material free of charge, and also lower prices on DVD courses for sale. Nevertheless, none of these competitors have had any success because they have not been able to compete with Moderskeppet's pedagogy.

Looked upon from this point of departure, museums ought to identify their own competitive strengths. However, let us discuss *Moderskeppet's* strengths in a museum context. Public museums are not commercial and should have no problems offering the public information and knowledge free of charge on their websites. The challenge is perhaps likely to be the pedagogical, and communicative, qualities. It is crucial that the sender, via the website, addresses readers in a proper and effective way. The answer to these demands is, however, not to engage professional communication agencies and to embed museums' communication within an impersonal agency discourse. What is striking about Moderskeppet's way of addressing their users is the complete absence of management and marketing language. The necessary technical language is furthermore simplified in material aimed for newcomers, and more professional for the experienced. Moderskeppet's communicative efficiency is a question of both content and form.

5.1.5 Less is more in Moderskeppet's website design

Pedagogical proficiencies also come through in the design of the website. The classical and harmonious horizontal direction constitutes an easily comprehensible architecture (see appendix). In the upper left corner, where we usually begin reading, the logotype attracts the eye. Immediately under the logo a row of buttons leads the visitor to free content. Another row of buttons, for the more frequent user, appears up to the right. Below the area with buttons and labels, visitors find frequently updated news to the left, general information about Photoshop and Moderskeppet framed to the right and beneath, marketing for Moderskeppet's new products, free as well as for sale. The news, information and marketing texts are all accompanied by visual communication. To view and read the bottom level one has to scroll once down the page. The design and amount of information follows the less is more principle. A website is, due to the hypertext format, possible to load with limitless amounts of information. Moderskeppet have decided to disregard the user's social demand because information overload might lead to dysfunction. Only the most necessary information is displayed and certain carefully chosen links appear. Neither is the website designed to meet the informational or functional demands of employees – this is rather a task for intranets. These principles might, however, be challenging for museums, as they are institutions with a culture of collecting and documenting in as extensive and detailed a manner as possible. Nevertheless, a public website should never be designed to meet the demands of the website's employees – or artefacts – but ought first and foremost to meet social demands among users, i.e. the museum's target groups.

5.1.6 Appropriation of digital communicative applications

From the very beginning Moderskeppet's website had the character of an irregularly updated booklet, not unlike many museum websites. This type of silent media only reach out to the most interested users, or to casual visitors looking for an institution's opening hours. To attract attention a website has to talk. In Moderskeppet's case the introduction of a daily blog became the communicative breakthrough. Within three weeks of the first posting, Moderskeppet's blog was mentioned in the blogosphere – the blog made Moderskeppet visible, listened to and talked about. Consequently, visitor frequency increased. In order to benefit from this, Moderskeppet launched a newsletter. By making people subscribe to the newsletter, a more direct and personal relationship was established with engaged users. In combination with the daily blog postings, written as if they were addressing every user in person, and newsletters with information for the "chosen" ones, a consumer community (Muniz and O'Guinn) was established around the Moderskeppet brand.

This ought to be instructive for museums. Museums employ people educated within the academic tradition of writing papers, essays and theses. They also employ people with aesthetic skills. Museums furthermore have important, interesting and appealing stories to tell. There is thus a message as well as people able to communicate the message. We can experience how well most museums succeed with this task when visiting their physical locations. Nevertheless, this demands performative more than communicative skills. Once an exhibition is produced the audience first and foremost consumes what is represented before them, in a one-way direction, by the museum's expertise.

A website, on the other hand, is a communicative space, demanding communicative skills and continuous communication. The term 'communication' is important, because websites can also involve visitors using interactive means. A communicative website provides the museum with limitless opportunities to make exhibitions and collections active and lively and to involve the audience in different ways. However, accomplishing daily compelling communication and interaction with users demands not only communicative skills, it demands time. Not spare time, not leisure time, but time included in employees' working hours. Museums with the vision to digitalise their operations beyond the website-as-booklet, must realise that such a decision demands that resources must be directed to daily communication within, and updating of, this space, once the infrastructure is put to work.

5.2 Involving users

Museums' overarching strategies and ideas actually make up important starting points for the analysis of user involvement as they frame museums' overall activities, including their views of both digital and IRL users. In methodological terms this part of the research consists of interviews with museum management and analyses of relevant documents relating to their view of users: Who are digital users (primary, secondary)? What are management's views of their digital users? Analysis of this empirical material will work on several levels. Management will of course analyse the material at the level of manifest content, but also work more critically in order to illuminate additional – implicit – dimensions in their views of users (Olsson).

Starting from these analyses, the next empirical step will be to identify social and cultural groups of users that might be specifically interesting to include in the study: perhaps the local community? Could these groups include young people? Or perhaps tourists? The selection of user categories should be made in cooperation with the museum's management, as the analyses of user preferences neither can nor should strive to construct generalised statistical knowledge of users. Hence, it is vital to make a strategic, informed selection of user groups to which to pay specific attention.

The third empirical step is to conduct focus group interviews (cf. Aljas and Pruulmann-Vengerfeldt). These focus groups need to cover at least four overarching themes: What are the participants' general views of museums? Do they have additional wishes in terms of what museums could offer them? Do they have specific wishes regarding museums' digital interfaces? And what forms of digital involvement with museums would they consider useful? This empirical material will also be analysed at different levels, both in terms of its manifest content and on the level of implicit assumptions and reflections concerning, for instance, the interviewees' views of museum digital applications.

6. Final reflections

The focus group interviews will constitute a starting point for the involvement of users in the development of regional museums' digital interfaces. The analyses of these interviews will inform the construction of digital interfaces – and practices – and might even serve as a corrective for those that already exist. Together with the analyses of best practice in the area of digital inclusion, they will hopefully add up to a solid state of knowledge for the development of museum digital interfaces. This developmental work is obviously in no way finished; it has merely begun. However, it has begun in a way that manages to involve the public – as citizens – from the very outset.

References

Aljas, Agnes, and Pille Pruulmann-Vengerfeldt. "Digital Cultural Heritage: Challenging Museums, Archives and Users." *Communicative Approaches to Politics and Ethics in Europe. The Intellectual Work of the 2009 European Media and Communication Doctoral Summer School.* Eds. Nico Carpentier et al. Tartu: University of Tartu Press, 2009. 59-71. Print.

Armstrong, Gary, and Philip Kotler. *Marketing: An Introduction.* 9th ed. Upper Saddle River, NJ: Prentice Hall, 2009. Print.

Ang, Ien. *Desperately Seeking the Audience.* London: Routledge, 1991. Print.

Barney, Darin. *The Network Society.* Oxford: Polity Press, 2004. Print.

Beiner, Ronald, ed. *Theorizing Citizenship.* Albany, NY: State University of New York Press, 1995. Print.

Business Dictionary. Vijay Luthra and BusinessDictionary.com, 2007-2011. Web. 4 Mar. 2012. http://www.businessdictionary.com.

Buskqvist, Ulf. "Interaktivitet." *Mediesamhället: Centrala begrepp* [Media society: essential concepts]. Eds. Peter Berglez, Ulrika Olausson. Lund: Studentlitteratur, 2009. Print.

Cameron, Fiona, and Sarah Kenderdine, eds. *Theorizing Digital Cultural Heritage: A Critical Discourse.* Cambridge, MA: MIT Press, 2007. Print.

Codrington, Stephen. "Applying the Concept of 'Best Practice' to International Schools." *Journal of Research in International Education* 3.2 (2004): 173-188. Print.

Costello, Anna B., and Jason W. Osborne. "Best Practices in Exploratory Factor Analysis: Four Recommendations for Getting the Most From Your Analysis." *Practical Assessment, Research and Evaluation* 10.7 (2005): 1-9. Web. 3 Mar. 2012. http://pareonline.net/pdf/v10n7.pdf.

Dahlgren, Peter. *Media and Political Engagement: Citizens, Communication, and Democracy.* New York: Cambridge University Press, 2009. Print.

Dana, Lucio E., and Kosmas X. Smyrnios. "Family Business Best Practices: Where from and Where to?" *Journal of Family Business Strategy* 1 (2010): 40-53. Print.

DuFour, Richard, and Robert Eaker. *Professional Learning Communities at Work™: Best Practices for Enhancing Student Achievement.* Bloomington, IN: National Educational Service, 1998. Print.

Grol, Richard, and Jeremy Grimshaw. "From Best Evidence to Best Practice: Effective Implementation of Change in Patients' Care." *The Lancet* 362.9391 (2003): 1225-1230. Print.

Habermas, Jürgen. *Structural Transformation of the Public Sphere.* Cambridge: Polity Press, 1989. Print.

Isin, Engin F., and Bryan S. Turner, eds. *Handbook of Citizenship Studies.* London: Sage, 2002. Print.

Jensen, Jens F. "'Interactivity': Tracking a New Concept in Media and Communication Studies." *Nordicom Review* 19.1 (1998): 185-204. Print.

Keene, Suzanne. *Digital Collections: Museums and the Information Age.* Oxford: Butterworth-Heinemann, 1998. Print.

Lang, Caroline, John Reeve, and Vicky Woollard, eds. *The Responsive Museum: Working with Audiences in the Twenty-First Century.* Aldershot: Ashgate, 2006. Print.

Marshall, T. H., and Tom Bottomore. *Citizenship and Social Class.* [1950] London: Pluto Press, 1992. Print.

Moe, Hallvard. "Dissemination and Dialogue in the Public Sphere: A Case for Public Service Media Online." *Media, Culture and Society* 30.3 (2008): 319-336. Print.

Muniz, Albert M., Jr, and Thomas C. O'Guinn. "Brand Community." *Journal of Consumer Research* 27.4 (2001): 412-432. Print.

Murdock, Graham. "Building the Digital Commons: Public Broadcasting in the Age of the Internet." The 2004 Spry Memorial Lecture. Vancouver, 18 Nov. 2004/ Montreal, 22 Nov. 2004. Web. 1 Apr. 2010. http://www.com.umontreal.ca/spry/old/spry-gm-lec.htm.

Olsson, Tobias. "For Activists, for Potential Voters, for Consumers: Three Modes of Producing the Civic Web." *Journal of Youth Studies* 11.5 (2008): 497-512. Print.

Purcell, John. "Best Practice and Best Fit: Chimera or cul-de-sac?" *Human Resource Management Journal* 9.3 (1999): 26-41. Print.

Raboy, Mark, and Bram Dov Abramson. "Grasping an Enigma – Media Policy and Social Demand." *Cultural Policy* 4.2 (1998): 329-355. Print.

Raboy, Mark, Bram Dov Abramson, Serge Proulx, and Roxanne Welters. "Media Policy, Audiences, and Social Demand: Research at the Interface of Policy Studies and Audience Studies." *Television and New Media* 2.2 (2001): 95-115. Print.

Regeringens proposition 2009/10:3. *Tid för kultur* [Time for culture]. Regeringskansliet. 21 Sep. 2009. Web. 1 Apr. 2010. http://www.regeringen.se/content/1/c6/13/21/04/a7e858d4.pdf.

Riksantikvarieämbetet. The Swedish National Heritage Board. Riksantikvarieämbetet. Web. 4 Mar. 2012. http://www.raa.se.

Smith, Colleen, and Frances Sutton. "Best Practice: What it is and what it is not." *International Journal of Nursing Practice* 5.2 (1999): 100-105. Print.

SOU 2009:15. *Kraftsamling! – museisamverkan ger resultat* [Mobilisation – cooperation between museums makes results]. Regeringskansliet. 16 Feb. 2009. Web. 1 Apr. 2010. http://www.regeringen.se/content/1/c6/12/06/93/70907ba4.pdf.

Stevenson, Nick. *Understanding Media Cultures: Social Theory and Mass Communication.* London: Sage, 1995. Print.

Webster, Frank. *Theories of the Information Society.* 2nd ed. London: Routledge, 2002. Print.

Appendix

Moderskeppets website (saved from Internet: 22 November 2010)

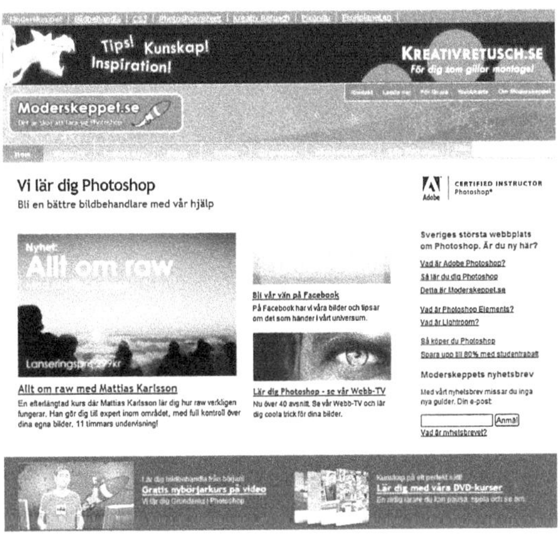

Tobias Olsson, PhD, is Professor of Media and Communication Studies at Lund University, Sweden. He has extensive research experience within the areas of media and citizenship, internet culture and mediated participation. His most recent publications include articles in *Javnost – The Public, Continuum: Journal of Media and Cultural Studies*, and Television and New Media. He is also the editor of a recent edited book titled "Producing the Internet" (Nordicom, 2013).

Anders Svensson, PhD, Assistant Professor of Media and Communication Studies at Jönköping University, and researcher within the national research project 'Organised Producers of Young Net Cultures'. His research interests include new media and democracy, citizenship and civic cultures, online communities and social capital, and critical analyses of participation and social media.

Changing Users of Memory Institutions

Krista Lepik

PhD Student
University of Tartu
Librarian
University of Tartu Library
Struve 1, 50091 Tartu, Estonia
krista.lepik@ut.ee

1. Introduction

In 1822, Auguste Comte was positive about inventing a strategic plan of scientific work to help society on restoring stability, harmony, and eventually, prospering. Coming to consensus, for illuminated men of his time, couldn't be an issue, wouldn't it? To this day, "social physics" has been developing, bringing about numerous theories and debates on society, concerning its different aspects and problems. Trying to describe and analyse another order: post-modern society, post-industrial society, information society, risk society, global society etc., numerous novel opportunities and threats are articulated. There is still a commonality that binds all different theories: it is encompassing changes or transformations, if you like to call it that way. Looks like never ever the change has been so rapid, ground-shaking, advancing or global. It fact, all these different theories actually try to grasp just one world, our life-world. Therefore, attempting to take into account major changes that have formed society and its inhabitants in the digital age, this article aims to explore and analyse transformations in memory institutions that could also be meaningful for the users of the latter. It should be done with precaution though, as it is complicated to establish surely what "new ideas, values or processes are genuine turning points in social history" (Bell 8).

In order to avoid confusion, some of the primary concepts need to be explained. No doubt that sometimes the concept of "memory institutions" can remain fuzzy (as at least some people would end up, talking about facilities to treat Alzheimer). In fact it is a so-called umbrella term for "libraries, museums, archives, heritage (monuments and sites) institutions, and aquaria and arboreta, zoological and botanical gardens" (Hjerppe 173). The list is not finite though as concerning documentation processes, journals and the systems of primary, secondary and tertiary literature, source literature and "repackaged literature" (e.g. textbooks) can also be considered as "memory institutions" (Hjørland 36).

It should be stressed that overarching concepts like "memory institutions" or "cultural heritage institutions" (also including museums, libraries, archives etc) are not something fixed but are also in transformation just like the institutions and phenomena they attempt to contain. As it is possible to see the term "memory institution" has already gradually changed in less than 10 years, depending on where *information* and therefore, *memory* is concealed. Similarly is "cultural heritage institution" also highly dependant on how we define *culture*, what is perceived as *heritage* and how these two concepts meet on institutional level. In a way it would be a bold yet naïve idea to attempt drawing conclusions on all different "memory institutions," putting them in a solid frame. The article is therefore focussed on museums and libraries but for further inquiry it can be obviously a challenge to analyse how transformations discussed here occur in other kinds of memory institutions as well.

Secondly, there is another fundamental question: how is a "user" defined and why is it *the* "user" we are talking about? In accordance with different activities performed in memory institutions, and various discourses having an impact on the latter, there are "visitors," "users," "patrons," "readers," "customers" etc. (Lepik, "Changing users" 273) memory institutions are created for. Of all those names or labels the "user" refers that, to certain extent, we are talking about "an active person, being able to search, choose, use, save and disseminate the information s/he is obtaining from the library" (ibid.). Compared to other names the concept of "user" is most neutral, though conveying perhaps a slightly positive connotation derived from "user-friendliness".

In a very broad sense, users of memory institutions can be both individuals and institutions. For the sake of clarity the *changing user* (of memory institution) is discussed at the level of individual user.

Methods applied for the article include concept mapping on the basis of different strategic documents and laws concerning directly the performance of University of Tartu Library (the statute of University of Tartu, the statute of University of Tartu Library, the obligatory copy law, the law of University of Tartu). 12 semi-structured expert interviews with faculty members and 7 with the personnel of University of Tartu Library have been conducted in 2009 and 2010. When analysing interviews concepts of critical discourse analysis were taken into account. Statistics from the University of Tartu Library's yearbook have been included as well.

2. The point of departure

Traditionally, memory institutions have functioned as guardians of cultural heritage, sharing "an over-arching common purpose of acquiring, organizing, preserving and presenting the cultural and scientific heritage of society" (Allen and Bishoff 59). Although dependant on overall openness of the society, memory institutions have acted as facilitators of cultural participation since the beginning of modern era: period when first public lending libraries and public museums were emerging, and first archives were made open for publics. Ever since memory institutions have been providing publics the access to written accounts of culture and science, proof of achievements of past heroes, and to what was perhaps considered as fundamental: the truth.

Although the article focuses on changes of digital age, it is not possible to underestimate earlier transformations in memory institutions. Today, these changes might not seem as rapid as contemporary ones but to a large extent the earlier changes have gradually shifted us to the point where a lot in memory institutions is taken for granted. It is essential that museums before digital era did not act merely as monuments to past times but they also have had a strong educative function (Kjeldbaek 121). There is no doubt that most of readers of this article have been enrolled to museums in big groups as school-pupils. Similarly in libraries, the librarian must have been able to recommend books, keeping in mind educative purposes, for the good of reader's development (nevertheless, leaving the choice of reading materials up to the borrower) (Sibul 8). From time to time though, authorities (especially in totalitarian regimes) have attempted to benefit from the educative function of museums and libraries, using them as tools distributing *right* ideologies (Johani 304), while archives have been locked tighter to hide *wrong* ones.

Memory as such is not something fixed or organised. The same applies for the documents and objects that have found their way into memory institutions. In order to get some sense of these entities, different classification and categorising systems were developed and eventually applied in halls full of card cabinets. In pre-digital era, understanding of these systems was metaphorically *the key* entrusted to our guardians of cultural heritage. Knowing *the key* (as it is an intangible notion it cannot be owned) has traditionally been the main feature of professionals in memory institutions and distinguishing value of the latter in society.

3. Transforming memory institutions

Approximately ten years ago, at the turn of the millennium, scientific literature in the field of information sciences was impregnated with exclamations about technological progress in memory institutions and anything that concerned these institutions: publications, databases, customer services etc. The excitement was understandable, as indeed, technological shifts in memory institutions have been most notable and tangible: at first, by shifting the cataloguing and reference work from "paper and pencil" and filing cabinets "to computers," and later to the Internet. Traditional functions of memory institutions: collecting, preserving and distributing documents/objects containing primary information and especially from now on, providing secondary information, were reinforced, emphasising once again the "key" role of memory institutions in digital age. The concept of ""“knowledge" workers (programmers, financiers, marketeers, designers, administrators) within an increasingly *service*-oriented labour force (where "servicing" includes a great variety of casual and menial work tasks as well as well-paid ones – the former often being undertaken by women)" (Hall, Held, and McLennan 7) seemed to fit well into new settings where professionals of memory institutions had found themselves in past decades.

Novel situation raised new questions and issues that all together have formed a *digital divide* in making available the metadata about collections of memory institutions. Financial issues can naturally be among first barriers that prevent from digitally describing the documents and later on, causing considerable problems in digitising politics. Secondly, especially in the case of museums, the object can have very specific characteristics and hence a question is raised: how to compress data about objects to uniform records. Thirdly, there can be an issue of cultural heritage as a cultural treasure – what if the information is available on the Internet and tempts someone to steal it?

If the developments above can be described as means of production in information society, there has to be a way to market the information product or service, hence a massive wave of marketing discourse has swept into memory institutions. On the basis of similar institutions (like museum among other museums, library among other libraries etc), memory institutions have found a consensus, treating each other as partners rather than competitors, attempting to find different ways of collaboration in order to strengthen their positions in society, and in the market as well. Though, the competition as such can take place in very different areas: scientific libraries as former legitimate providers of scientific information for their communities are in a way competing with databases, publishers, media monitoring firms etc. Although museums possess valuable information on cultural heritage, they have competitors in leisure industry as well: among concert halls, restaurants, sightseeing objects etc. The situation for public libraries can be similar to the one of museums – from the aspect of providing amusing activities for local community.

An issue of the extent of following marketing rules in memory institutions became hence an important question. From a university library (where some research has been conducted), parallels can be drawn to other memory institutions. On one hand, they should be active "service providers that pays attention to different target groups" and "act as a meeting place rather than sanctuary". But there's also *the status*. In a case of the university library, it may happen, as one of the faculty members claimed, that "if the library is seen among faculty as a customer service provider rather than equal partner with mutual goals, this concept may start spreading also on the university as a whole as in some sense the students perceive the university library and university as an integral whole" (academic staff) (Lepik, "Tartu University Library" 6). Something very similar may happen in other memory institutions where, because of attracting and amusing the customers (not "visitors" anymore), the function or value of these institutions can be interpreted by their users differently from the interpretation of professionals. If everything goes so quickly and simply in memory institutions, aren't they eventually resembling to supermarkets where (information) needs can be satisfied, or a leisure service consumed in a simple and quick way?

When following basic marketing rules, the question of target groups is traditionally posed. Due to the global nature of Internet the user of memory institution's digital application could literally be anyone in the world. Therefore, mapping of target groups becomes more complicated and the same applies for providing customised services. This problem is solvable from technical perspective but in this case, another issue, concerning privacy of the user, is raising: even if "we *would* have more sophisticated tools for analysing the profile of user of the digital library, where would the thin red line be between researching and "spying"?" (Lepik, "Changing Users" 279)

In addition, the concept of "stakeholdership" is also familiar to memory institutions – the people who act as stakeholders of the memory institutions are perhaps not always articulated as such but it is obvious that not all users of memory institutions are treated as consumers (of information, culture etc.). Take for example the symbiosis in universities, between the faculty members and subject librarians from university libraries. Or have a look at various communities that are called as "museum friends". In both settings, there are people who are formally "outsiders" of particular memory institutions but who nevertheless contribute a big portion of their time and knowledge for the good of this institution, hence obtaining an honourable position of a stakeholder.

Another transformation, also tightly attached to widespread of Internet access, has taken place in the field of distance work (including distance learning). Studies (Christiansen et al.; Anthony; Lepik, "Changing Users" etc.) reveal that in case somebody just needs the access to information that is a traditional function of all memory institutions, the digital version is often preferred to non-digital one. Memory institutions are visited rather in case the necessary information is not digitised, due to specific information needs or for other purposes than retrieving documents. Memory institutions are still functioning as learning spaces (probably due to developments in the ways pupils and students are taught at schools or universities) but very often the aspect of learning (as acquisition of new information) is also paired with leisure activities, contributing to overall personal fulfilment and satisfaction (Falk).

As a magic wand, the concept of Web 2.0 has come into play during recent years. It brings along transformations in the field of information technology, tapping to the notion of "participation," instead "publishing" (O'Reilly). At the same time it also provides a new option, also for memory institutions as well, to develop public relations and marketing. In a way, almost every transformation mapped above has to more or less extent affected or been affected by Web 2.0 applications.

The most intangible yet present transformation in memory institution is probably the one that is tied to expertness. In contemporary society, memory institutions can be seen as experts who are at critical points expected to provide the members of society the answers about past related to the present (Giddens, "Living in a" 63), to create "a shallow screen that intervenes between our present lives and our history" (Thompson 269). So, even if professionals in memory institutions are not always treated as part of new skill-based elites (Bell 361) they are still, as preservers of our *collective memory*, considered to be highly credible especially for people who are now responsible for their own Giddensian "reflexive project of the self" ("Modernity and Self-Identity"), hence developing their own biographies. What makes the situation confusing though is that in some cases, memory institutions are actually ready to attribute some of their intrinsic expertness to their users, by stepping back and "letting the object speak for itself" (Kjeldbaek 124) or even, enabling users to curate exhibitions themselves.

4. Discussion

Transformations in memory institutions are strongly shaped by the chicken or the egg causality dilemma. It is typical that, as one of the most notable changes, the technological ones are paid attention first. But we might ask why it is so? Why do we take superiority of technical changes in memory institutions so much for granted? Instead, it could be necessity that drives memory institutions to develop newer and better applications for their users. Necessity of being nice to customers and therefore coping well at the information market. Necessity of being up-to-date in rapidly developing (information) society. Necessity for an institution that literally would help to separate the grain from the chaff in (reflexive) society. And so forth, the list provided hereby does not claim to be comprehensive as memory institutions' roles in society can be really very diverse. Nevertheless, in order to be satisfied all these needs still ask for tools and this is where technical applications come into play.

In this article the concept of "changing users" (of memory institutions) are understood as a polysemantic concept. It can refer to a user whose identity, societal roles (and own perceptions of roles), needs, values etc are in transformation. It can also point to a user the memory institutions are attempting to alter in a more or less implicit or even explicit manner. The transformations explained in this article are of the latter nature but a large part of the following discussion is dedicated to tension between these two aspects of change.

It is debatable whether different transformations (to make using of memory institution more convenient, faster, more simple etc.) are really in the interest of users or is it just about somewhat worried hanging behind societal trends (and legitimising the existence in novel situations). It could also be the users who let memory institutions know about their needs directly or indirectly, thereby initiating some sort of transformations (for example, "voting with their feet" and going to spend some time in a gallery instead of museum or working at home or office, using digital library's applications instead of a physical library building). Summing up: are new roles and tasks appointed (in this case there is a question "by whom?") to memory institutions or are they actively choosing and taking these tasks themselves?

When paying attention to the users it is also worth to ask to what extent they are attracted by different societal transformations. Perhaps they are not interested in these at all and just wish to find some peace and quiet in memory institutions or, unconditional distribution of memory. It appears that the silence (explicitly present in so-called 1st generation museums (Kjeldbaek) and in other memory institutions of that time) helps provide a refuge and escape from personal-social pressures. The latter enables tension release, slowing down mentally and escaping from role overloads (Falk 49) – eventually providing some *recharging* and *rejuvenating* effects (ibid. 69). We can only assume though that this tendency is connected with appreciation of memory institutions as experts.

This takes us to a classical question about the extent of intervention and encouragement to participation in civil society. In fact, different authors have shifted this extent back and forth again: from bearers of more or less socialist mindset who prefer interfering, to liberals proposing *lassez-faire* attitude. Quite recently, Jakob Nielsen has pointed out a 90:9:1 scheme on participation inequality in online environments, whereas 90% of users are "lurkers" (i.e. read or observe but don't contribute), 9% of users are intermittent users contributing from time to time, and 1% of users participate a lot and account for most of contributions. What if the same applies in other democratising settings, including memory institutions as well? Is it in the nature of people or is it just about some hints and tricks to increase the extent of cultural participation in somewhat untraditional settings such as memory institutions?

Timing and essence of transformations can highly vary in different memory institutions. Some changes (like enabling more participation) may arrive earlier to, for example, museums as there these are more urgent or more relevant than in other memory institutions. Hence, the transformations of particular memory institution's role play an important role: whether the main trend is to enable leisure activities (Falk), create cultural biography (van Mensch 18) or please world-class academics. In each case, especially if the funds are scarce, choices are made and the impact to diverse developments' paces is also different.

It is obvious that different transformations encompass diverse roles for both memory institutions and their users. Hence another issue is emerging: are memory institutions evaluating these changes as users do or is there any kind of divide or tension in understanding of the changes (for example, what changes demand quicker reacting, what if another societal issue is more burning but instead, memory institutions are being supported etc.)? In case there is some congruity in role perceptions and memory institutions are treating users as customers and the latter also perceive themselves as customers it can take both to beneficial relationship. But if the roles are perceived differently: for example, if users are treated as customers but the latter are willing to act as stakeholders, contributing to well-being of the institutions, the tension is about to come into being.

As mentioned above, this article cannot embrace the diversity of societal transformations in digital age. So, side by side with these questions posed in discussion, a variety of other issues could be dealt as well. For instance, what kind of impact would shifts in demographical processes have on memory institutions? Or what kind of effects would environmental issues have to transformation of memory institutions and their users (take for example, devastating actions of hurricanes in America or floods in Europe or Asia)? As noted in the introduction, all these changes take place in a single space: in our life-world. So perhaps being aware of the nature of transformations and its relations to processes in society helps both memory institutions and their users to face new challenges in digital age.

Therefore, almost century-old words of wisdom should be kept in mind: "[this] rapid change of conditions affords an *opportunity* for progress, but is not itself progress" (Dewey 312). The transformations in digital age might not always appear as progressive, but it depends on memory institutions how or to what extent to react in order to cope with or, even better, to benefit from new challenges.

5. Conclusion

In contemporary society there are various transformations that pose novel settings for both memory institutions and their users. The most tangible one is, of course, the technological change. Nevertheless, it is coupled with issues of changing role-models, renewing market positions and power relations in and around memory institutions. Side by side with newly emerging developments the users still tend to remain veiled by their secrets. The mystery of the user is strongly attached to memory institutions fate: will future generations find the memory institutions as valuable (or at least, useful) as they are now, or what kind of challenges will memory institutions face in next 10, 25 or 50 years.

Acknowledgements

This article has been prepared with the help of Estonian Science Fund grants no. 8006 and no. 7162.

References

Allen, Nancy, and Liz Bishoff. "Academic Library/Museum Collaboration: I'm OK, You're OK." *ACRL X Conference Proceedings*. Chicago: Association of College and Research Libraries, 2001: 59-69. Web. 23 Sep. 2010.

Anthony, Kristin. "Reconnecting the Disconnects: Library Outreach to Faculty as Addressed in the Literature." *College and Undergraduate Libraries* 17.1 (2010): 79-92. Web. 1 Mar. 2010.

Bell, Daniel. *The Coming of Post-Industrial Society: A Venture in Social Forecasting*. New York: Basic Books, 1999. Print.

Christiansen, Lars, Mindy Stombler, and Lyn Thaxton. "A Report on Library-Faculty Relations from a Sociological Perspective." *The Journal of Academic Librarianship* 30.2 (2004): 116-121.Web. 1 Mar. 2010.

Dewey, John. "Progress." *The International Journal of Ethics* 26.3 (1916): 311-322. Print.

Falk, John H. *Identity and the Museum Visitor Experience*. Walnut Creek, CA: Left Coast Press, 2009. Print.

Giddens, Anthony. *Modernity and Self-Identity: Self and Society in the Late Modern Age*. Cambridge: Polity Press, 1991. Print.

—. "Living in a Post-Traditional Society." *Reflexive Modernization: Politics, Tradition and Aesthetics in the Modern Social Order.* Eds. Ulrich Beck, Anthony Giddens and Scott Lash. Cambridge: Polity Press, 1994. 56-109. Print.

Hall, Stuart, David Held, and Gregor McLennan. "Introduction." *Modernity and its Futures.* Eds. Stuart Hall, David Held and Tony McGrew. Cambridge: Polity Press in association with Open University, 2003. 1-11. Print.

Hjerppe, Roland. "A Framework for the Description of Generalized Documents." *Advances in Knowledge Organization* 4 (1994): 173-180. Print.

Hjørland, Birger. "Documents, Memory Institutions and Information Science." *Journal of Documentation.* 56.1 (2000): 27-41. Web. 23 Sep. 2010.

Johani, Helene, ed. *Raamatukoguhoidja käsiraamat.* Tallinn: Eesti Riiklik Kirjastus, 1964. Print.

Kjeldbaek, Esben. "Post-Modernism and the Three Generations of Museums." *Nordisk Museologi* 1-2 (2001): 119-126. Print.

Lepik, Krista. "Changing Users of the Digital Library." *Transforming Culture in the Digital Age. International Conference in Tartu 14-16 April 2010.* Ed. Agnes Aljas et al. Tartu: Estonian National Museum, Estonian Literary Museum, University of Tartu, 2010. 273-280. Web. 28 Sep. 2010.

—. "Tartu University Library as a Medium between Different Patrons." *Qualitative and Quantitative Methods in Libraries: Theory and Application. Proceedings of the International Conference on QQML2009, 26-29 May 2009, Chania, Crete, Greece.* Eds. Anthi Katsirikou and Christos H. Skiadas. 2009. Web. 1 Mar. 2010.

Nielsen, Jakob. "Participation Inequality: Encouraging More Users to Contribute." *Jakob Nielsen's Alertbox.* 9 Oct. 2006. Web. 11 Nov. 2010.

O'Reilly, Tim. *What is Web 2.0: Design Patterns and Business Models for the Next Generation of Software.* Web. 10 Nov. 2010.

Sibul, Aleksander. *Raamatukoguhoidja käsiraamat.* Tartu: Noor-Eesti Kirjastus, 1926. Print.

Thompson, Kenneth. "Social Pluralism and Post-Modernity." *Modernity and its Futures.* Ed. Stuart Hall, David Held and Tony McGrew. Cambridge: Polity Press in association with Open University, 2003. 221-271. Print.

Van Mensch, Peter. "Annotating the Environment. Heritage and New Technologies." *Nordisk Museologi* 2 (2005): 17-27. Print.

Krista Lepik is a PhD Student in the Institute of Journalism and Communication, Faculty of Social Sciences and Education, University of Tartu, Estonia. She also works at the University of Tartu Library. Her research interests include cultural participation issues in (Estonian) memory institutions.

Audiovisual Collections in a Digital Culture: Reflections on Providers and Users of Digital Audiovisual Heritage in Flanders

Lien Mostmans

Eva Van Passel

Researchers
IBBT-SMIT, Vrije Universiteit Brussel
Pleinlaan 9
Brussels, 1050
Belgium
lien.mostmans@vub.ac.be
eva.van.passel@vub.ac.be

1. Introduction

Many theorists have elaborated on an emerging digital culture, outlined its principal components or conceptualised the changing user roles within this digital cultural realm. Digital culture is in an ongoing state of transition. These transformation processes instigate interesting questions on the extent to which certain sectors are ready to embrace a far-reaching digital cultural evolution and the hybridisation of traditionally distinct roles this entails. In this article, we wish to look at the emergence of aspects of digital culture in one specific domain, i.e. the digital audiovisual heritage realm in Flanders (Belgium). Digital audiovisual heritage is by no means an established domain; the role of providers has not yet crystallised, and the low maturity of the field makes large-scale empirical user research far from straightforward. The way in which the often-theorised changing agent roles will evolve in practice within the digital audiovisual heritage field remains an important area of research. The aim of this article is to look at whether preliminary empirical results on the digital audiovisual heritage field in Flanders hint at the emergence of digital culture. The first section will sketch the wider context, looking at audiovisual collections on the one hand and the wider realm of digital culture on the other hand. The second section will focus on the Flemish research project BOM-VL (Archiving and Distribution of Multimedia in Flanders). The main aim of this article is to revisit some of the principal component notions of digital culture in light of the project's findings.

2. Context

2.1 Audiovisual collections as an integral part of a society's heritage

The research project BOM-VL, which we will discuss in more detail further in this article, is motivated by the fact that the important heritage value of audiovisual collections has been widely recognised. UNESCO's *Recommendation for the Safeguarding and Preservation of Moving Images*, adopted in 1980, exemplifies this (UNESCO 229). The *Recommendation* aims to address the threat to large parts of society's audiovisual heritage caused by the vulnerability of analogue collections. Audiovisual heritage can be defined in a number of ways (Edmondson 21-26). UNESCO's approach is very inclusive on the one hand, including also "non-literary or graphical material, such as photographs, maps, manuscripts, slides and other visual works, selected in their own right" (ibid. 21), but on the other hand the aforementioned *Recommendation* stresses that moving images are perhaps the ones that are most vulnerable, in contrast to for example photographs. The focus of this article will therefore be on the more narrow definition of audiovisual heritage as 'moving images' in particular, a focus largely shared by the BOM-VL project. This vulnerability includes issues related to the carriers of audiovisual content, such as chemical decomposition (cellulose nitrate film) or 'vinegar syndrome' (triacetate film), but also the rapid obsolescence of the technologies needed to play back these carriers (ibid. 44-47). More recently, this threat can be seen as one of the causes of the increase in the number of initiatives concerned with the digitisation of audiovisual collections and materials. On the one hand, digitisation is seen as addressing the need for preservation of valuable audiovisual heritage; on the other hand, digitisation can also greatly increase the possibilities of access to audiovisual content. Future access does not only require preservation, but it is also often stated that preservation cannot be seen as an end in itself but must always have future access in mind (ibid. 19). Moreover, it is important to realise that digitisation cannot be deemed the sole solution in light of the preservation of audiovisual content: "It is only slightly facetious to say that digital information lasts forever – or five years, whichever comes first" (Rothenberg 42). Digital vulnerability concerns the deterioration of digital media and data carriers, but also the obsolescence of software and hardware (see also Rosenzweig). While digitisation in its narrowest sense – the process of encoding analogue source material into digital files – may offer a temporary solution to the threat of analogue obsolescence, long-term policies must go beyond this narrow digitisation scope and tackle the issue of digital sustainability. This is an enormous challenge for institutions that hold audiovisual collections and has been one of the main underlying motivations of the BOM-VL project.

2.2 Digital culture: Key concepts and user roles

2.2.1 Participation, interaction and changing user roles in a digital culture

The increasing digitisation of audiovisual collections as outlined above is embedded within a larger context, i.e. the emergence of digital culture in more general terms. In the next section, a selected set of concepts will be highlighted in order to portray digital culture. Before doing so, we already wish to address one underlying aspect of this changing cultural landscape: the role of the end user, consumer or visitor within this digital culture on the one hand, the role of the provider or producer of digital materials on the other hand, and finally the interaction between and hybridisation of these formerly more distinct roles. Many authors and scholars have acknowledged the emergence of a more participatory culture (Jenkins, "Convergence Culture"; Benkler; Manovich, "The Language of New Media"). Moreover, the blurring boundaries between user and producer roles have instigated numerous different conceptualisations of a 'new user'. Prosumer (Toffler), produser (Bruns, "Gatewatching, Not Gatekeeping"; Bruns, "Gatewatching") and Pro-Am (Leadbeater and Miller) are some of the terms used to describe these types of (former) users and the ever-increasingly hybrid roles they can take on. As hybrid concepts – respectively combining producer and consumer, producer and user and professional and amateur – these neologisms all hint at a more active type of user and at the convergence between user and producer roles. The aforementioned concepts are not necessarily introduced within the framework of theories on digital culture specifically. The concept of produser, for example, has been introduced by Bruns in the realm of online news; however, it has consequently also been applied to describe developments in other areas of the online world (Bruns, "Blogs, Wikipedia, Second Life and Beyond"). The concept of the Pro-Am can be found throughout a range of activities people might engage in, as is illustrated by the numerous examples given by Leadbeater and Miller. For the remainder of this article, the term produser will be preferred.

The tendency of blurring boundaries between user and producer roles is highly relevant in light of the realm of digital information. It is interesting to assess whether the case of digital audiovisual heritage hints at these new hybrid user roles. Regardless of the context in which the aforementioned neologisms describing these roles have emerged, the concepts are indeed useful within the domain of digital audiovisual heritage as well; they might in fact be paramount in understanding the process of digitally interacting with heritage. On the one hand, it seems straightforward to state that the user as well as the provider of digital audiovisual heritage materials can play an important role in the process of creating meaning through participation in the heritage field and interaction with heritage materials. On the other hand, in light of the presented theories, it is not unthinkable that their roles might merge altogether, creating true produsers of digital heritage collections. Upon discussing some principal components of digital culture, this hybridisation of roles always has to be kept in mind.

2.2.2 Taking part in a digital participatory culture: A selection of concepts

In the following paragraphs we will present a select literature overview in order to contribute to further understandings of digital culture, its constituent notions and characterisations. We will specifically discuss the notions of bricolage and remediation, as examples of manifest expressions of an environment in which traditional user and provider roles are evolving and (online) (digital) audiovisual heritage fields are ever more employing a bottom-up approach (Deuze, "Participation, Remediation, Bricolage"; Deuze, "Convergence Culture in the Creative Industries"; Hand; Jenkins, "The Cultural Logic of Media Convergence"; Manovich, "Remixability"). Moreover, this environment is generally characterised as highly participatory and interactive. We therefore perceive participation and interaction as meta-components and as backgrounds against which users and providers handle audiovisual heritage in a digital culture.

Definitions of interactivity not only appear to be rather dispersed throughout different research domains, they also tend to be discordant. It is indeed referred to as a multi-discursive concept in that it "can be found with significantly different meanings or connotations according to their use within different discourses" (O'Sullivan et al., qtd. in Jensen 188). Moreover, Jensen distinguishes interaction from interactivity, in that interaction refers to actions of two or more individuals observed to be mutually interdependent, but not to mediated communication, and interactivity refers to media use and mediated communication (Jensen 200). Conversely, Kiousis does not explicitly distinguish both notions. Both mentioned authors show, however, that interaction and interactivity alike are often studied and described as possible categorisations of technology, as Jensen's own deducted interpretation demonstrates as well, as "a measure of a media's [*sic*] potential ability to let the user exert an influence on the content and/or form of the mediated communication" (201). And although Kiousis' definition of interaction is expressed in sociological terms, with an emphasis on communication between two or more people, he describes interactivity as a degree to which communication technology can create a mediated environment in which participants can communicate (one-to-one, one-to-many, and many-to-many), both synchronously and asynchronously, participate in reciprocal message exchange, and perceive the experience as a simulation of interpersonal experience (Kiousis 372). Conversely, we are inclined to opt for a broad understanding of interaction. Whereas both authors focus on interaction via or with a *medium* as a *means* of communication, we want to consider interaction via or with *content* not only as a communicative act but also as a *goal* in itself.

Participation is an equally characterising feature of contemporary digital practices, and like interaction it depicts an environment that enables a certain engagement with materials at hand. Based on a literature review, Deuze primarily discusses participation in a news production setting with emergent participatory news and 'we media' practices, alternative and citizens' media and in a more political context with a shift from a rather passive informational citizenship to monitorial and voluntaristic informational citizenship in a glocalised community. In this context, participation commonly refers to a "culture of participatory authorship" (Deuze, "Participation, Remediation, Bricolage" 68) with what we interpret as a growing expectation, opportunity and ability to intervene and modify mediated (news) content. Moreover, participation has been discussed in other frameworks as well, such as political contexts (e.g. Carpentier) or marketing research studies (e.g. Van Oost).

Against the background of interaction and participation, two specific practices have surfaced in the digital content realm: remediation and bricolage. Replicating Bolter and Grusin, Deuze argues that remediation refers to the effects of new media on older media, causing old media to 'refashion' themselves, or otherwise "a remediation of old media by new media" (Deuze, "Participation, Remediation, Bricolage" 69) and embeds it in the same news production environment as mentioned above. In another context, Elkins (qtd. in Bowen 220) exposes a dichotomy between new and old forms of art production and the manner in which visual artists are able to include both manual and digital processes, or operate digitally all together. We would therefore argue for a broad understanding and examination of the remediation concept, not in the least with a specific attention for creative activities in a digital environment. Furthermore, Deuze ("Participation, Remediation, Bricolage") deals with bricolage, which has been described in the past in media and cultural studies as a creative practice in which existing artefacts are partly or entirely appropriated, incorporated and re-used when creating objects (Hartley 22-24). Strongly tied in with the idea of originality, bricolage practices are frequently involved in contemporary issues relating to plagiarism and copyright. Although the author situates this component in a broader perspective than the previous ones, for example by citing Chandler's vision on bricolage by means of creative undertakings on personal homepages, the focal point is on news operation practices. More recently, other scholars have aptly illustrated the 'return of the bricoleur' in the wider media landscape (e.g. Frissen and Slot).

3. Digital audiovisual heritage and the case of BOM-VL

3.1 BOM-VL: Aim and methodology

The research project BOM-VL[1] ('Bewaring en Ontsluiting van Multimediale data in Vlaanderen', the Dutch acronym for Archiving and Distribution of Multimedia in Flanders) forms an apt empirical background to confront the more theoretical principal components of digital culture as outlined above, and to look at how the digital audiovisual heritage field in Flanders might take shape within this culture in transition. The project aimed to find sustainable solutions for obstacles to the processes of archiving, preserving and distributing digital audiovisual content originating from the cultural sector as well as audiovisual broadcasters in Flanders. As stated in the opening section of this article, both preservation of and access to audiovisual heritage need to be ensured; BOM-VL acknowledged this dual need in its archiving and distribution focus respectively. Problems of this nature had never been addressed on a large scale and in such a comprehensive multidisciplinary fashion in Flanders before. Problems addressed within the project were manifold and included technical, legal and organisational issues related to digital preservation and access. This article will address research within two specific trajectories, i.e. user (group) requirements related to digital access and architectures for digital archiving and distribution. The trajectory in which user (group) requirements were examined and mapped will be explained in more detail upon readdressing the concept of bricolage. The research results with regards to distribution modalities will mainly be elaborated on in the section on remediation. While the digitisation of society in a more general sense forms the wider context in which the digitisation of the audiovisual heritage field takes place, digital culture as outlined in the previous section had not been the main focus of BOM-VL. In this article, we revisit the findings in order to establish whether they hint at the existence of some of the principal components of digital culture in practice, most notably the practices of bricolage and remediation.

Because user requirements and user and provider roles are central in this article, it is indispensable to indicate that research results in the discussed strands

1 The project BOM-VL (January 2008 – June 2009) was made possible by the support of the Flemish Community and was carried out by a consortium of partners from the Flemish audiovisual, broadcast and cultural sectors and several IBBT research centres. IBBT (Interdisciplinary institute for Broadband Technology) is a research institute founded by the Flemish Government. A full list of project partners and more details on IBBT-SMIT's contribution can be found on http://smit.vub.ac.be/project/25/BOM-VL. All research reports, the majority in Dutch, have been made available for download on the project website https://projects.ibbt.be/bom-vl/.

did not include user research: the field was still crystallising, as were its prospective users. From a methodological point of view, the discussed potential user requirements were mapped through literature research and case studies on the one hand and a panel session involving experts on the other hand. Distribution modalities were researched by means of a domain analysis and expert interviews. More specifically, by means of in-depth expert interviews within three stipulated target groups in the BOM-VL project, i.e. the educational sector, the audiovisual creative industry and cultural institutions, different distribution models within the different sub-sectors were mapped and their viability was assessed. In this article we particularly draw on the interviews with experts from cultural institutions and discuss their views on the distribution of digital audiovisual content in the cultural heritage sector. In both the user requirement trajectory and the distribution modalities trajectory, the experts were selected based on their domain-specific expertise, the position they hold in the field and their ability to transcend sectoral visions and issues. The individual interviews were conducted in the period February-March 2009 and were built on semi-structured interview schedules in order to maximise the comparability and increase the reliability and credibility of data whereas a number of questions and sub-questions were presented less rigidly as discussion topics. The expert panel discussion was conducted in June 2008 and was structured around a discussion text and a number of discussion statements provided to the panellists beforehand. All interviews and discussions were fully transcribed, coded and analysed by means of a pre-structured analytic schedule.

3.2 BOM-VL: Findings revisited in light of digital culture

3.2.1 Bricolage from a BOM-VL perspective

Within the BOM-VL user requirements trajectory, four approaches to an audiovisual archive were distinguished for research, each entailing specific user needs. The division was made according to type of use rather than user group in acknowledgement of the possibility that certain users might have diverging needs depending on the context of use. Two specific research topics are perhaps the main areas invoking the concept of bricolage and the user as bricoleur: (1) audiovisual archive material as a source for creative processes; (2) archives that are open and dynamic to some extent. Firstly, the conceptual elaboration in the research report on archives as a source for creativity explicitly mentions bricolage and audiovisual bricoleurs (Debuysere 41, 52, 68, 76). Associated terms addressed by Debuysere include collage, found footage cinema, sampling, détournement, remix, cut-ups and mashups. The study shows that the Flemish bricoleur does indeed exist: some practices of Flemish artists who can be described as bricoleurs have been provided as case study examples (ibid. 59-64). However, additional findings from the BOM-VL project, i.e. the findings related to user requirements with regards to open and dynamic archives (Van Passel and Beyl), might nuance the current scope of bricolage.

As the BOM-VL project did not include large-scale user research, this strand followed the assumption that at least a proportion of users will expect some requirements of open and dynamic archives to be met, for example to make connections among dispersed media content, annotate content or interact with media producers in dynamic ways. This assumption is based on theories on participatory digital culture (e.g. Jenkins, "Convergence Culture"; Benkler; Deuze, "Participation, Remediation, Bricolage"); moreover, it is illustrated by numerous online examples and Web 2.0 developments. The requirements for an archive to be categorised as open and dynamic were established by means of a literature study and a preliminary domain analysis. This was followed by more in-depth case studies of Web 2.0 initiatives and research projects selected to illustrate aspects of open and dynamic archives. In order to establish potential recommendations with regard to implementing these aspects in BOM-VL and beyond, an expert panel discussion was organised. The selected Flemish experts' combined expertise, practical as well as academic, could be situated in the fields of cultural (heritage) content, online cultural initiatives, Web 2.0 and user-generated metadata.

One particular finding from this expert panel discussion is highly relevant for this article: while the core assumption that at least a proportion of users would expect to be able to interact dynamically with archives was acknowledged, large question marks were raised when it comes to the extent of this proportion in Flanders. The experts pointed to the potential lack of a critical mass of active users or produsers, which might lower the priority of several aspects of open and dynamic archives. Following their comments, we can perhaps assume that the realm of digital audiovisual content use in Flanders is as yet not characterised by bricolage to a large degree. It is therefore unclear to what extent "rapidly growing numbers of ordinary users manifest themselves again as professional amateurs, tinkering with the digital building blocks that are available to them in the rich treasury of materials and experiences that the web offers" (Frissen and Slot 96). As a result, the digital audiovisual content realm might still have a way to go before it can be characterised as a true materialisation of digital culture.

3.2.2 Remediation from a BOM-VL perspective

Remediation not only applies to old media, as is often emphasised; authenticity concerns (e.g. the loss of valuable and relevant data), quality-related issues (e.g. unsatisfactory image quality as a result of compression) and the obsolescence of players and file formats (causing an increasing migration of data) underline the continuous need to refashion digital media as well. In order to distribute collections digitally (online) to target audiences in a meaningful way, collection managers also need to take into account their particular needs and expectations, and possibly refashion previous approaches. Conversely, the BOM-VL project has shown that bringing this into practice is not evident. One research trajectory focused on architectures for digital preservation and distribution and included interviews with representatives in the educational sector, the audiovisual creative industry and cultural institutions (Mostmans et al.). Our results indicate that there are not only significant dissimilarities in user requirements but also in requirements from a provider's perspective, taking into account that providers can also be users. Given the limited number of interviews, certain prudence is called for when interpreting the results. However, they can be considered as careful indications of different sectoral attitudes vis-à-vis certain distribution models and modalities.

An important condition for using and distributing audiovisual materials, emphasised by representatives in the education sector, can be reduced to didactics. Stakeholders in this sector have questioned a merely instrumental or illustrative use of audiovisual materials in classrooms and advocated a broader implementation of audiovisual materials, for example with an emphasis on visual education. This approach requires a certain adaptation or remediation of audiovisual content: audiovisual material needs to be more extensively contextualised, even to the point where fully digital lessons can be made available. Although this assumes a certain commitment from teachers, syllabus makers, and/or other intermediaries, the interviews have revealed that remediating basic metadata into more elaborate user-generated metadata can form a threshold for teachers. Such involvement appears rather premature in the education field in Flanders. Conversely, even though opinions on distribution models diverged significantly, informants in the cultural sector generally considered distribution models as raw tools, which can advance the enrichment of self-generated content of an audiovisual nature. Extensive contextualisation of audiovisual content, as is deemed required in educational fields, was not mentioned as a prime modality. Stakeholders in the audiovisual creative industry underline that metadata and broader forms of contextualisation can facilitate a better organisation of collections and increase findability.

To restate, taking into account changing user requirements and accordingly remediating digital material, as the discussion on the integration of audiovisual materials in classroom didactics has showed, are paramount to developing distribution strategies for audiovisual platforms. The issues in the project are indicative for a broader concern when dealing with disclosing and distributing audiovisual archives. It has become clear that a homogenous approach is not satisfactory; analogue and digital materials require flexible target group specific and continuous remediation in order to maximise their reach and value.

4. Discussion

Our research indicates that remediation, both from a provider's perspective (how to annotate and present collections to a target audience?) and a user's perspective (how to give meaning to rough archive material?) in a digital culture is not as evident as the ever-increasing development of user-centred or Web 2.0 technologies in certain digital and online environments might suggest. While the BOM-VL project has certainly acknowledged notions of participatory culture at its outset, it has equally demonstrated that the current field and the stakeholders involved might inhibit the fluent emergence of an interactive and participatory digital audiovisual heritage domain. The section on bricolage has stipulated that the proportion of Flemish users wishing to interact with audiovisual source material or wishing to engage in produsage activities is likely to be relatively small. While digital availability of audiovisual materials might facilitate participation and interaction for some produsers, this does not entail that digital audiovisual heritage will automatically lead to an interactive and participatory culture. This is not only true for (prod)users, but also from a provider's perspective: remediating content and metadata towards target user groups is more labour intensive than just throwing digital materials out there and waiting to see what happens. Thus, while it is clear that digital tools can facilitate user-centred changes, it might be too forthcoming to assume that user and providers alike will automatically adapt their roles to technological evolutions.

In addition to requiring motivation and certain capabilities from users as well as providers, participatory ways of interacting with audiovisual heritage also presuppose a critical mass of digital data and metadata. This is an additional vital issue in Flanders, demonstrated by the results of the assessment of the current digitisation progress in cultural (heritage) institutions and broadcasters in Flanders (Nulens and Debuysere). BOM-VL's recently completed follow-up project Archipel[2] (a network-centric approach to sustainable digital archives) has continued to address

2 Archipel (October 2009 – December 2011) was a project funded by IWT (Agency for Innovation by Science and Technology, Flanders - http://www.iwt.be/) in the PIM framework (Programme Innovative Media). A full list of project partners, including IBBT research centres, can be found on the project website http://www.archipel-project.be/.

the need for sustainable archiving and distribution solutions. Within Archipel, it has been important to acknowledge that a true emergence of digital culture in Flanders can still be questioned. Gaining insight into the needs of user groups remains an issue, e.g. within the strand where sustainable valorisation models for digital archiving and distribution have been developed, mainly focusing on the target groups education, research and cultural industries.

This article has therefore worked towards identifying important areas for future research. A first recommended topic stems from the initiation of BOM-VL. As it did not include large-scale empirical user research, it would be interesting to establish to what extent users of or visitors to digital audiovisual heritage collections can be deemed active users, produsers, remediators or bricoleurs. As the digital audiovisual heritage domain is still in transition in Flanders, it would be useful to revisit it at a later stage, in order to examine to what extent the principal components of digital culture will have permeated the audiovisual heritage realm once it has reached a higher maturity level. Secondly, the degree to which providers will continue to give meaning to audiovisual heritage collections needs to remain the focus of continuing attention. This has been partially achieved within the recently completed research project EPICS[3] (E-learning Platforms in the Cultural heritage Sector), which has focused on contextualising and distributing audiovisual content for educational purposes.

After having looked at interacting with and giving meaning to audiovisual heritage in the digital era, many questions remain. Will user and provider roles merge into one hybrid category of produsers, and to what extent? What does being a produser entail in the digital heritage realm? Which proportion of visitors to online audiovisual heritage in Flanders can be described as produsers? It is not inconceivable that the distinction between the roles of user and provider will become obsolete in the future. However, in an era already often conceptualised as a culture in which these roles are increasingly converging, findings suggest that interaction, participation, remediation and bricolage can still be problematic, both for users and providers.

3 EPICS (November 2009 – October 2011) was a project co-funded by IBBT. More information about the EPICS consortium and its goals can be found on the project's blog http://ibbtepics.wordpress.com (in Dutch).

References

Benkler, Yochai. *The Wealth of Networks: How Social Production Transforms Markets and Freedom.* New Haven; London: Yale University Press, 2006. Print.

Bolter, Jay David, and Richard Grusin. *Remediation: Understanding New Media.* Cambridge, MA: MIT Press, 1999. Print.

Bowen, Tracey. "Making Art in a Digital/cyber Culture: Exploring the Dialectic between the Manual Creator and the Digital Self." *Digital Creativity* 14.4 (2003): 219-227. Print.

Bruns, Axel. "Gatewatching, Not Gatekeeping: Collaborative Online News." *Media International Australia* 107 (2003): 31-44. Web. 17 June 2012.

—. *Gatewatching: Collaborative Online News Production.* New York: Peter Lang, 2005. Print.

—. *Blogs, Wikipedia, Second Life, and Beyond: From Production to Produsage.* New York: Peter Lang, 2008. Print.

Carpentier, Nico. "Theorie 1: Het TIP-model. Een analystisch kader voor culturele participatie." *Cultuur en ICT. Verkennende analyses en gevalstudies 2002-2003.* Eds. Nico Carpentier, Bart Cammaerts, and Olga Van Oost. Gent: Steunpunt Re-Creatief Vlaanderen, 2003. 55-61. Web. 17 June 2012.

Chandler, Daniel. "Personal Homepages and the Construction of Identities on the Web." University of Wales, Aberystwyth. 9-11 September 1998. Paper presentation at Aberystwyth Post-International Group Conference on Linking Theory and Practice: Issues in the Politics of Identity. Web. 17 June 2012.

Debuysere, Stoffel. *Werkpakket 1: Gebruikersnoden. Deeltaak 2: Creatief Atelier – cultuursector.* Flanders: BOM-VL Consortium, 2008. Web. 17 June 2012.

Deuze, Mark. "Participation, Remediation, Bricolage: Considering Principal Components of a Digital Culture." *The Information Society* 22.2 (2006): 63-75. Web. 17 June 2012.

—. "Convergence Culture in the Creative Industries." *International Journal of Cultural Studies* 10.2 (2007): 243-263. Web. 17 June 2012.

Edmondson, Ray. *Audiovisual Archiving: Philosophy and Principles.* Paris: UNESCO, 2004. Web. 17 June 2012.

Frissen, Valerie, and Mijke Slot. "The Return of the Bricoleur: Redefining Media Business." *The Good, the Bad and the Challenging: The User and the Future of Information and Communication Technologies: Conference proceedings. Copenhagen, Denmark, 13th-15th May 2009.* Eds. Bartolomeo Sapio, Leslie Haddon, Enid Mante-Meijer, Leopoldina Fortunati, Tomaž Turk, and Eugene Loos. Koper: ABS Center, 2009. 88-99. Print.

Hand, Martin. *Making Digital Cultures: Access, Interactivity, and Authenticity.* Aldershot: Ashgate, 2008. Print.

Hartley, John. *Communication, Cultural and Media Studies: The Key Concepts*. New York; London: Routledge, 2002. Print.

Jenkins, Henry. "The Cultural Logic of Media Convergence." *International Journal of Cultural Studies* 7.1 (2004): 33-43. Print.

—. *Convergence Culture: Where Old and New Media Collide*. New York; London New York University Press, 2006. Print.

Jensen, Jens F. "'Interactivity': Tracking a New Concept in Media and Communication Studies." *Nordicom Review* 19.1 (1998): 185-204. Web. 17 June 2012.

Kiousis, Spiro. "Interactivity: A Concept Explication." *New Media & Society* 4.3 (2002): 355-383. Web. 17 June 2012.

Leadbeater, Charles, and Paul Miller. *The Pro-Am Revolution: How Enthusiasts Are Changing Our Economy and Society*. London: Demos, 2004. Web. 17 June 2012.

Manovich, Lev. *The Language of New Media*. Cambridge, MA; London: MIT Press, 2001. Web. 17 June 2012.

—. "Remixability: Remixing and Remixability." *Manovich.net*, 2005. Web. 1 Mar. 2010. www.manovich.net/DOCS/Remix_modular.doc.

Mostmans, Lien, Eva Van Passel, Stijn.Bannier, and Gert Nulens. *Werkpakket 5: Architectuur digitale bewaring en ontsluiting. Deeltaak 5.3: Aanbevelingen doelgroepscenario's*. Flanders: BOM-VL, 2009. Web. 17 June 2012.

Nulens, Gert, and Stoffel Debuysere, eds. *Werkpakket 5: Architectuur digitale bewaring en ontsluiting. Deeltaak 5.1: Aanbodmodaliteiten*. Flanders: BOM-VL, 2008. Web. 17 June 2012.

O'Sullivan, Tim, John Hartley, Danny Saunders, Martin Montgomery, and John Fiske. *Key Concepts in Communication and Cultural Studies*. London; New York: Routledge, 1994. Web. 17 June 2012.

Rosenzweig, Roy. "Scarcity or Abundance? Preserving the Past in a Digital Era." *American Historical Review* 108.3 (2003): 735-762. Web. 17 June 2012.

Rothenberg, Jeff. "Ensuring the Longevity of Digital Documents." *Scientific American* 272.1 (1995): 42-47. Web. 17 June 2012.

Toffler, Alvin. *The Third Wave. The Classic Study of Tomorrow*. New York: Bantam Books, 1980. Print.

UNESCO "Recommendation for the Safeguarding and Preservation of Moving Images. Adapted by the General Conference at its Twenty-first Session, Belgrade, 27 October 1980." *Conventions and Recommendations of Unesco Concerning the Protection of the Cultural Heritage*. Paris: UNESCO (1983): 225-239. Web. 17 June 2012.

Van Oost, O. "Verkenning 2: Museum Participation in a Network Society: A Reflection." *Cultuur en ICT. Verkennende analyses en gevalstudies 2002-2003.* Eds. Nico Carpentier, Bart Cammaerts, and Olga Van Oost,. Gent: Steunpunt Re-Creatief Vlaanderen (2003): 36-54. Web. 17 June 2012.

Van Passel, Eva, and Joke Beyl. *Werkpakket 1: Gebruikersnoden. Deeltaak 1: Open en dynamisch archief.* Flanders: BOM-VL Consortium, 2008. Web. 17 June 2012.

Lien Mostmans worked at IBBT-SMIT, Vrije Universiteit Brussel (Belgium) from September 2008 until April 2012. She has been involved in several research projects concerning culture in a networked society, including projects on digital preservation and distribution of audiovisual content, on e-learning for cultural heritage and on young people's cultural experiences.

Eva Van Passel joined IBBT-SMIT, Vrije Universiteit Brussel (Belgium) in November 2007. Her research focuses on arts and culture in a networked society, with recent research projects mainly centring on organisational, financial and policy challenges related to digitisation, digital preservation and access to audiovisual content and cultural heritage.

Heritage, User and the Digital Environment: Rewriting the Narrative of the Literary Past

Marin Laak

Senior Researcher
Estonian Literary Museum
Estonian Cultural Historical Archives
Vanemuise 42
51003 Tartu
Estonia
marin@kirmus.ee

1. Introduction

Cultural transformation processes and the new digital media have influenced the creation and consumption of culture, but have also opened up new directions for memory institutions – museums, archives and libraries preserving traces of the past. Memory institutions are facing the task of digitising their collections of cultural heritage, and related to this task, a lot of principal problems have emerged. For instance, when these institutions start creating digital collections, they first have to find an answer to the main question: how to make a selection among millions of memory objects and how to represent these objects? The main idea of my article is to explain the exploitation possibilities of cultural heritage in the new digital media environment. My second aim is to introduce cultural heritage users' experiences and to expand the notion of the 'user' according to different practices.

The notion of the 'user' is in the foreground of contemporary new media discussions and is closely related to the theory of participatory culture (Jenkins). The social network channels of new media communication, as well as the needs of the youngest user groups, have been brought into the limelight. In their article "Digital Cultural Heritage – Challenging Museums, Archives and Users" Pille Pruulmann-Vengerfeldt and Agnes Aljas have, from the point of view of new media communication, stressed the importance of participatory culture in the context of the activities of memory institutions as well. The article states that the third main task of memory institutions, besides preserving and introducing the cultural heritage collections, is to take into consideration the principles of participatory culture in their routine work. The authors are convinced that the third basic task of memory institutions is to invite audiences "to become active participants in introducing, learning and being involved with cultural heritage" (Pruulmann-Vengerfeldt, Aljas 111). This statement links the traditional activities of memory institutions (collection, preservation and distribution of cultural heritage) with participatory culture and new media user requirements. For memory institutions, these new issues could pose an intriguing challenge because of complicated and resource-consuming digitisation processes.

In this article I am going to explain these processes by considering the practices of memory institutions and user habits, examining the users' other activities besides media consumption, on-line participation and social networking and give three different examples of the use of cultural heritage. These examples or "usage practices" are related to the development of content-based environments, as well as to the accompanying theoretical reflections. I wish to relate the notion of the 'user' with the theoretical constructions of remediating cultural heritage and the historiography of cultural history narratives in the new media environment. For the introduction I am going to give a short overview of Internet usage focusing on the needs of the younger generation, and the user group of professionals working at memory institutions. Second, I will introduce the main concepts, theoretical problems and results of three application projects as cultural heritage usage practices.

The empirical basis for my paper is the Estonian Literary Museum (ELM) – a memory institution engaged in collecting and preserving national cultural heritage, and managing huge archives of Estonian culture, but also a national research and development centre, aimed at the study of cultural history using its archival sources. The Estonian Cultural History Archives of the ELM is a central institution storing archival materials concerning cultural history (collections of manuscripts, correspondence, photographs, art, audio and video recordings, memories and life stories). The Estonian Folklore Archives of the ELM is a central archive storing Estonian and Finno-Ugric folklore (collections of manuscripts, photographs, sound and video recordings), and the Archival Library of the ELM is a central library storing all Estonian printed documents (books, periodicals, pamphlets, maps and bookplates). This is the rich and semantically meaningful environment that has had an effect on the subject and focus of the current paper. During the past 15 years, my research has centred on the construction of three non-linear models for presenting cultural history in the new media environment, based on the objects of cultural heritage held in the different archives of the ELM. These models have been created as 'laboratories' within which to discuss the representation of our cultural past in the digital space.

Assuming that the notion of literary history includes the broad notion of cultural history and, being at the same time a part of subjective cultural memory, we can pose a more specific question: for whom and how should we construct and remediate the cultural past and cultural memory in the new digital media environment? In this paper I am going to introduce some Estonian experiences in investigating user habits.

2. Activities of the new media 'user'

Gunther Kress develops further the concept of the user, arguing that the use of existing resources without changing them has to be replaced by remaking them in the process of use, in different acts and interaction (Kress "Visual and Verbal Models" 74). Such a change in the recipient's activities is caused by the multimodal character of new media that transforms words, images, sounds and moving images into one and the same text and thereby, unites the forms of production (writing, drawing, designing) and reception (reading, listening, watching, learning) that have so far existed separately. The term of writing is transformed: instead of conveying causal relations in the form of a linear text, it now means a spatial combination of, mainly, verbal and visual texts. Kress writes that two aspects have gained importance – an increase in visuality and the replacement of narration with representation and description (see also Livingstone 221-224). The new writing – "reconfiguration of writing" – is based on hypertext as the paradigm of a new critical thinking that involves the ability to summon different kinds of information in order to solve problems. Hypertext requires an active reader and, at the same time, it creates such a reader. The authors of hypertext can link information and create paths through a corpus of texts to annotate existing texts, add references and bibliographic data (Landow "The Paradigm is More Important Than the Purchase" 41-42). It has very often been said that "the reader becomes the author of the work". It has most often been argued that through the choices the reader/user always has to make while moving in the digital environment, he/she becomes the co-creator of the content – the writer. The characteristics of new media that enable rapid communication between the creator of the content (the author of the text) and the reader/user, helping him/her to influence the content, have also been pointed out.

Constructing cultural knowledge in the web-environments we should take into consideration the fact that memory institutions holding and preserving cultural heritage – the archives, museums and libraries – can be seen as the mediators of cultural memory as well as the users of this heritage.[1] Who are the poten-

1 See also the article by Lien Mostmans and Eva Van Passel, "Audiovisual Collections in a Digital Culture: Reflections on Providers and Users of Digital Audiovisual Heritage in Flanders" in this publication.

tial producers and users of the digital collections at the memory institutions, what kind of habits do they have? To answer these questions, we can get help from the results of sociological surveys.

2.1 Quantitative nationwide survey

The users' habits of new media consumption in Estonia were studied in the representative set of nation-wide surveys called *Mina. Maailm. Meedia* ('Me. The World. The Media') carried out by the Media and Communication Institute of the University of Tartu. The surveys were conducted in 2002, 2005 and 2008, based on a sample that consisted of 1,500 respondents in the age range 15 to 74. The results enable us to make generalisations about current Internet usage practices. We can see that over six years the level of Internet usage increased in every user group. By 2008 it had reached almost 100% among the youngest population groups, and doubled among the middle-aged and older groups. The average level among the entire population of Estonia at the end of 2008 was 70% (Aljas, Pruulmann-Vengerfeldt 117-119; Pruulmann-Vengerfeldt, Aljas 112-115; Laak, Pruulmann-Vengerfeldt "Literary Past as the Cultural Memory", and "Re-Writing Literary Past" 168). Figure 1 gives the data from the 'Me. The World. The Media' national survey carried out by the Institute of Journalism and Communication at the University of Tartu in autumn 2008:

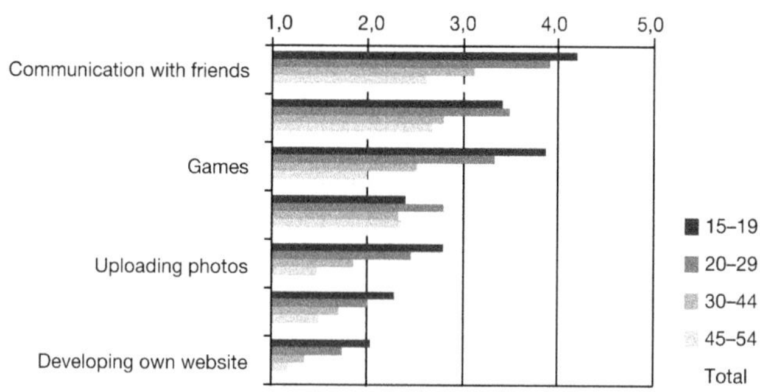

Figure 1: *Internet user types according to user purpose: the averages of user assessments on the 5-point scale (1 unimportant, up to 5 very important). Source: Survey "Me. The World. The Media", 2008*

Using cluster analysis, Pille Pruulmann-Vengerfeldt and Agnes Aljas (113-114) outlined six basic Internet user types according to their usage purposes: database usage, entertainment, social networks, work and study, virtual arts and uploading photos. Remarkable age differences can be seen here. For the youngest groups, work and study and also entertainment are equally important, but they are actively contributing to online content as well: uploading photos and participating in social networking sites. For them, the Internet has become a multifunctional environment. The middle-aged user group is just the opposite, mainly work-oriented and focused on practical activities. Their online communication is considerably non-personal. Qualitative interviews with the specialists of memory institutions (34 people) and focus groups of high school and university students (20 people) confirmed the data of the large surveys.

2.2 Qualitative survey: The focus group interviews

In order to analyse the expectations of the young user group, as part of the "Cultural Heritage in the Digital Age: Problems of Transformation and Receptions" research grant we conducted two focus group interviews with high school and university students in 2008 (see Pruulmann-Vengerfeldt, Aljas 116-122). The results confirmed the data from the surveys and added details about the needs and expectations of young users relating to the web pages of memory institutions.

We asked the recipients to describe an ideal web environment of a memory institution. From their discussions, 5 main principles can be summarised.

1. The physical experience of seeing the originals in a museum is clearly not so important for the youngest groups; digital collections should be sustained as online representations, e.g.: "A memory institution must have a presence on the Web along with all of its content, as it is often not possible for users to visit the institution".
2. Secondly, we can see the concept of various user levels; all the database users are not equally able to give their own interpretations of the objects, e.g.: "Multifaceted information should be structured pursuant to the user profiles, so that it would be possible to distinguish between information that is relevant for researchers and meant for users who just try to find interesting things".
3. Other principles were related to semantic relations between the digitised objects, e.g.: "A database must contain an introduction to its structure and data, abundant illustrations, video materials and interviews", or "The data (list of sources, digitised sources, etc.) must, in addition to links/references to other related databases, also include interpretations, contexts and background information that would help create associations".

4. The super-database of all cultural heritage materials seems to be a common dream for all, e.g.: "Various cultural heritage databases should be consolidated in one environment and the structure should be unified".

The young generation, which we often call the new media generation, has grown up with digital technologies. For them, the first and the fourth principles are the most relevant. The memory institutions should be ready for mediation of the cultural heritage to this new multimedia generation (Laak, Pruulmann-Vengerfeldt "Re-Writing Literary Past" 232-233).

According to Pille Pruulmann-Vengerfeldt and Agnes Aljas, the key focus of the interviews was centred on making digitised materials available to users. They stress that:

> [t]his inevitably boils down to the question of maximally effective information architecture. With increasing amounts of information available online, both users and producers of online materials feel that the searchability, clarity and variety of information is vital. ... we can say that in many ways, the online spaces and databases of the museums and archives provide a multitude of challenges. The first role of digital cultural heritage is to aid the storage of artifacts and to save them for the future. (Pruulmann-Vengerfeldt, Aljas 124-125)

At the same time, the institutions should be interested in increasing the number of users and archive participatory possibilities and "therefore, grasp the potentials and opportunities hailed by new technologies" (ibid.) that provide ways to close the gap between memory institutions and the new generation. In the authors' opinion, unless there is a considerable change in the way memory institutions think about the audiences of heritage, this potential may never be realised.

3. Usage of digitised cultural heritage in cultural historiography

Literary historians of the Estonian Literary Museum and the University of Tartu have studied the problem of user groups to overcome the miscommunication between the users' different habits, mentioned by Pille Pruulmann-Vengerfeldt and Agnes Aljas (125). Pruulmann-Vengerfeldt and Aljas have constructed more than 10 years worth of content-based digital information spaces and non-linear models of literary history narrative. These activities were based on the new digital literary history writing and the statement that the traditional approach to the past as a linear 'story', based on the causal chain and the consequence of facts was replaced by a notion of literary history as a space-time filled with breaks and interruptions. Metaphors of 'labyrinth', 'network' and 'patterns' were often used when speaking of the architecture of the new kind of "building" of literary history narrative, and the architecture of literary history "building" was imagined as a gallery of many entrances already in the mid-1990s (see Rojola; Saariluoma). The digital era introduced possibilities for rewriting the literary history narrative in a new way and using new constructions, but the architecture itself preserved its main feature – it was an open-ended and semantically meaningful labyrinth that followed the pattern structure of *Garden of Forking Paths* by Jorge Luis Borges as analysed by Sophia Psarra in her article "'The Book and the Labyrinth Were One and the Same.' Narrative Architecture in Borges' Fictions" (86-87).

I started my research by first clarifying the opportunities, challenges and requirements presented to the re-writing of literary history narrative in the digital environment. To do this, we had to take into consideration the attributes of two different realms: 1) the components of traditional historiography of the literary history, 2) the qualities of the new media environment that could be productive in representing literary history, and 3) the activities of the new media 'user'.

3.1 Components of the traditional historiography of the literary history

The wider objectives concerning the investigation for opportunities to relate the older literary culture and the developing digital culture – in order to ensure the continuity of cultural memory – are: 1) to develop a model or models of literary history that are as universal as possible and can be applied in the digital environment, and 2) to clarify the possibilities of studying the literary texts and literary history that come into being in the digital environment. According to the case of Estonian cultural history writing we can present several generalisations, which should be taken into account in developing non-linear models in the digital environment.

First, we should consider that so far, literary histories had for centuries been dominated by traditions. For instance, the writing of Estonian literary history has a long tradition starting from the first decade of the 20th century, or even earlier, from the last decade of the 19th century.[2] A literary history narrative has traditionally been a subjective story by one or several authors and connected with continuously changing contexts, ideologies, interpretations, etc. These aspects have all revealed the need for periodic rewriting of literary histories, similar to other historical narratives.

2 The beginning of literary history writing in Estonia is problematic because the first treatments of Estonian literary history added any writings in Estonian (e.g. single words and phrases) starting from the 13th century, as well as oral tradition, to the notion 'literary history'. The first literary history in a narrower sense, *Kurze Geschichte der ehstnischen Literatur*, was published by Dietrich Heinrich Jürgenson in 1843-1844. The first contemporary literary history based on literary works and fiction, written by a member of Young Estonia movement, Gustav Suits, was published in *Die Kultur der Gegenwart* (Berlin-Leipzig) in 1908.

Proceeding from this tradition, we can see that the macro-elements of a literary history narrative include literary and aesthetic processes, the biography and bibliography of the author, literary works, literary scene, socio-cultural background, activities of literary institutions, literary reviews and criticism and visual materials (photos, book covers, autographs, etc.). We ask how to apply such macro-elements in the construction of the architecture of a digital narrative for remediating literary history in the digital space.

This aim was well supported by the Estonian memory institutions, whose priorities include, besides preserving cultural heritage, making this heritage accessible in digitised format. A difference in the work of a memory institution and the writing of literary history is pointed out – while a memory institution can be a warehouse of digital databases, the representation and remediation of the literary history narrative in the digital environment requires a content-based conception worked out according to the content of the materials (Laak "Kirjandusloo mittelineaarsed mudelid" 320).

Particular characteristics of new digital media (interactivity, multimediality, possibility to create and visualise the relations between the objects, unlimited space, etc.) have made it possible to remediate any linear narrative using the new tools. This challenge is relevant also to the new literary historiography. I should stress that we can conceive literary history as an integrated (poly)system and as an "umbrella" concept, and as "the idea of a model" (Even-Zohar 31), not just the reception of single texts from single authors. In this system, a literary text (a work) is only one media object among others. Using the new, digital environment we can represent the past through its objective, unchanging and authentic cultural heritage sources that have been preserved in the memory institutions (Laak "Kirjandusloo mittelineaarsed mudelid" 320, and "Beyond Digital Labyrinth" 48-50).

3.2 Qualities of the new media environment

Speaking about rewriting and remediating history, we should take into consideration the distinctive characteristics of new media and the web environment. In addition to the phenomenon of participatory culture as, for instance, in *Convergence culture* by Henry Jenkins, we should stress the second characteristic of the new media environment – the increasing visuality. The users of the electronic environment have naturally adopted visual, not verbal logic. Pictorial and visual representations have become central headwords. We could add that, similarity to other, illustrative materials, the screen is also treated as an image. Discussions about visual grammar and the semiotic organisation of the screen have proceeded from this point of view (see Kress; Livingstone).

The third distinctive characteristic treatment is the transformation of the notion of 'writing'. The digital era has opened up paths to a new type of text creation. Its most important feature is the breaking of the linearity of the text – the narrative proceeding from event to event and the causal reason-result chain – and its replacement with non-linear models of the text.

The fourth distinctive characteristic is interactivity, which has more often been mentioned as a characteristic of the new media environment. This means that the role of the reader becomes more complicated. Lev Manovich uses the concept of interactivity only because of its computer-specific qualities. He believes that computer technology presupposes a human-computer interface (HCI), containing 1) physical objects – monitor, keyboard, mouse, etc.; 2) the way of organising data on the screen, the use of icons; 3) manipulation of the data via dynamic menus, copying, cutting, etc. According to Manovich, an object will become interactive as soon as it appears on the screen. We have to differentiate between different types of computer technology-related interactive structures and operations, for instance, "closed" and "open" interactivity (55-57).

The concept of 'interactivity' initiates a discussion on the activities of the recipient in the new media environment. Compared with the recipient of the linear text, his/her activities have become more complex and multilayered, combining use, different traditional types of reception (reading, watching) and writing. Raine Koskimaa has thoroughly studied the activities of an interactive reader in the digital environment; his doctoral thesis *Digital Literature* is largely based on Espen Aarseth's idea of "ergodic" literature that emerges in the digital environment in *Cybertext: Perspectives on Ergodic Literature* (1997) that emphasises the connotations of the 'interactivity' and 'reader'/'user' problematics with earlier works. Koskimaa writes:

> 'Interactivity' is a highly problematic term in regard to literature /.../ – a fact which was systematically formulated by Roman Ingarden already in thirties in his work *Das Literarische Kunstwerk* (1960). This idea was further developed, especially, in reception aesthetics and reader-response studies; these research practices take it as their starting point that the reader is an active participant in the literary signification process. There are several kinds of interactivity, and the difference between the interactivity of conventional literature and that of digital literature was, for the first time, clearly expressed by Espen Aarseth, who decribes four categories of reader(/user) functions: interpretation, navigation, configuration, and writing.[3]

However, in the hypertextual environment, the reader acquires an additional need for/opportunity for rapid movement between different parts of the text, of configuring it through his/her choices and, at the last level, of complementing it by adding his/her own text and links. The activities of such a new type of user have two focuses, connected with multimodality and hypertextuality.

3 See http://users.jyu.fi/~koskimaa/thesis/chapter1.htm.

4. Constructing the architecture of the non-linear narrative of cultural history

Constructing non-linear narratives in the digital environment has been influenced by George B. Landow's praise of socio-cultural knowledge environments (*Post-colonial Web; Victorian Web; The Women Writers Project*) and hypertext theory in the 1990s (Landow "Hyper/Text/Theory" and "Hypertext 2.0"; see also Bolter 19-20).

4.1 ERNI – Estonian Literary History in Texts 1924 to 1925

We tested the possibility of using the hypertext model of literary history with the ERNI (Estonian Literary History in Texts 1924 to 1925)[4] pilot project between 1997 and 2001. This is a network-like model of literary history, based on the theoretical ideas of reception aesthetics. The conception of this model was mainly based on a hypertextual network of critical texts from one short period; it opened the literary history process by highlighting a synchronous reception level – interpretation by the literary critic as a reader. In practice, it embodies a large complex of tightly interlinked critical texts written in 1918 to 1926, and reflects the literary scene and the process of literary life of the 1920s. Biographical data of the authors has been added because of the need to explain the position of a particular author in the cultural field of the time.

Principally, ERNI does not strive to offer its users a ready-made long linear narrative. It contains a number of small stories that were visualised by intertextual threads as the new knowledge models of the literary past (Laak, Viires "Representing Canonical Texts"). Its reader/user is offered the opportunity to participate in the process of reception history and to write his/her personal interpretation of literary history, using different genres (see Laak, Viires "Digital Environment as a Constructor of Literary History"; see also Laak "Eesti kirjandusloo traditsioonist").

ERNI's network-like model was based on hypertextual relations; it showed that it is possible to reproduce reception history in the digital environment, while also pointing out the shortcomings of such a venture. Problems arise with the readers because in the late 1990s we had no experience in guiding the readers of such works. It was clear that hypertextual, network-like literary history will bring forth a new type of a reader – the interactive reader mentioned above, who can, by opening linked literary sources, create his/her 'own' literary history whenever it suits him/her, being free of interpretations forced by the author of literary history as a "linear narrative" printed in a book. ERNI left literary history principally

4 See http://www2.kirmus.ee/erni.

open for different interpretations. However, these interpretations raised the question of coherence of literary historiography as one type of memory writing and also questioned the meaning of the relations that came into being when the texts were linked.

We could say that coherence of texts is the key question of modelling a literary history in the digital environment. This causes the changes in the notion of 'writing' in a digital space mentioned in the first part of this article, acquiring the new meaning of creating and describing intertextual relations. All the following interpretations of literary historical facts, which allow us to talk about literary history as the history of interpretations, are secondary and can occur only after relations between literary macro-elements such as 'fractals' (e.g. biographical data, works, archival documents, reviews) have been made digitally available.

In the theory of new media, discussions on the reader are closely connected with the problems of the recipient (reader/user) and changes in his/her role. Such discussions more and more involve the teaching of new types of texts generated in the digital environment and a new field of digital pedagogy, recently also explored by Landow in his *Hypertext 3.0* (2006). Attention was drawn to the need to teach critical literacy and to describe the meaningful content of electronic links/relations already in the 1990s, when computers were applied in teaching.

The methods, pedagogy and the larger realm of e-learning based on digital content will form a broader issue that is in its initial stage in the context of the Estonian case based on ERNI. Today, such models as ERNI, which were constructed at the end of the 1990s, have become a little outdated. Hypertextual networks are being replaced by special software for the creation of large content environments where it is possible to construct the relations between the objects more systematically.

According to Gunter Kress, mentioned above, we can conclude that it is possible to analyse the activities of historians as users mediating archival sources to the wider public in the new media environment.

4.2 The intertextual model of the Estonian national epic Kalevipoeg

The problem of systematic presentation of the relations was one of the main issues in our second project *The Estonian National Epic Kalevipoeg*, carried out in the framework of the *CULTOS* project in 2001-2003[5]. Here, our goal was to develop, define and apply the ontology of intertextual relations of all kinds of texts (in the wider sense), and to create software that would allow the transport of these related texts. An international work group developed a comprehensive CULTOS Ontology of Intertextuality, which is also a database of relations formulated by computer

5 See http://www.cultos.org.

technological mapping of intertextual relations that stand above the discussion on the concept of intertextuality. In the context of the semantic web, this is an exhaustive ontology that describes the relations between all possible multimedia texts from cartoons, comics and texts of fiction up to pop songs, radio broadcasts and TV clips.

The other question is how the new technology influences the creation of knowledge? The theoretical considerations for the construction of the proposed model are connected with the text of the Estonian national epic *Kalevipoeg* (Kreutzwald; Laak "Beyond Digital Labyrinth" 57-59). The study of *Kalevipoeg* in the digital environment and by means of appropriate computer software required an approach that totally differed from the previous research tradition. In order to visualise intertextual relations, the full text of the epic (19,032 verses) had to be split into motifs and small segments. During this work we discovered that:

a) various motifs are gradationally repeated on the intra-textual level of the epic, and they start forming small independent stories, e.g. the Harp, the Virgin, the Sword, the Father, the Journey, Sleep, etc. (Laak, Viires "Intertextuality and Technology").
b) The second discovery was that *Kalevipoeg* was found to be intertextually related to archetexts of European culture, as well as with the newest textual levels of contemporary Estonian culture, especially poetry. We could claim that computer technology allows the representation of new types of cultural knowledge models – intertextual threads (Laak, Viires "Representing Canonical Texts").

4.3 The model of the Kreutzwald's Century

The experience of constructing the semantic threads of cultural sources has had a great influence on our next project *Kreutzwald's Century: Estonian Cultural History Web*,[6] started in 2004 and still in progress.

The *Kreutzwald's Century* project was intended as an interactive environment joining a new user-friendly interface and the content-based selections of cultural sources. For this project, we digitised the sources held at the archives and library of the ELM and also added a lot of supplementary information. The results of our eight-year-work and the complicated architecture of the Estonian cultural historical web can briefly be introduced with the help of different project menus.

6 See http://kreutzwald.kirmus.ee.

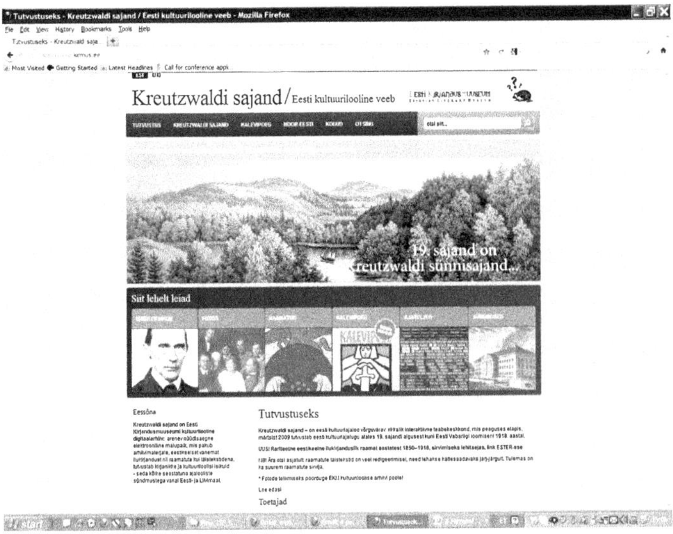

Figure 2. *The main page of* Kreutzwald's Century: The Estonian Cultural History Web – *"The 19th century is the century of Kreutzwald's birth ..."*

Kreutzwald's Century represented Estonian cultural history from the 19th century up to the beginning of the 20th century. With the help of five dynamic time axes we can see how Estonian written culture developed rapidly after the era of the Enlightenment and the Estonian national awakening. Canonical Estonian and European writers were introduced into the context of Estonian social and cultural history on five parallel temporal axes:

1. Kreutzwald's life and books – 82 events
2. Estonian socio-cultural history – 875 events
3. European literature and its canonical authors – the biographies of 119 authors
4. Estonian literature and its canonical authors – the biographies of 90 authors and the first editions of their books (image and text versions)
5. The story of the creation of *Kalevipoeg* – 44 events.

Digitised cultural sources have been linked with the biographical data of the writers. The user can read a writer's texts and also browse his or her books, manuscripts, documents and photograph collections.

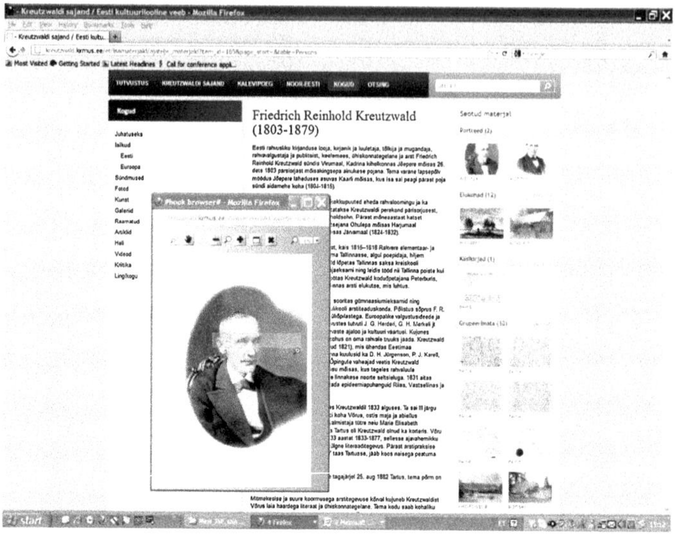

Figure 3: The main menu Collection (Isikud) of the archive of F. R. Kreutzwald (1803-1879), the author of Kalevipoeg.

The aim of the project was to construct a cultural historical space based on the macro elements of the cultural history narrative mentioned above.

4.4 The context of memory institutions

We could ask of how this project differs from other projects or digitisation practises in Estonian memory institutions because the digitisation of cultural heritage is not confined to two or three content-based projects of the Estonian Literary Museum. Other Estonian archives, libraries and museums have a number of other digitisation projects; some of them are related to the *Europeana* and other EU projects. The Estonian National Library, the University of Tartu Library, and the Archival Library of the ELM are cooperating in the largest current digitalisation projects in Estonia: *Red Book*, collecting the rarest Estonian books published since the 16th century,[7] and *DEA*, collecting the oldest newspapers from the 19th century.[8] These cultural heritage collections have been chosen and digitised according to the priorities of the memory institutions, e.g. usage, condition, and the intention of long-term preservation. The user can find the images of digitised books only if she/he knows the exact title or author of the work, publishing data, etc. If we compare digitisation

7 See http://www.nlib.ee/PunaneRaamat.

8 See http://dea.nlib.ee.

practices of the archives and libraries to our ideas of using the digitised sources for writing history in our content-based model of *Kreutzwald's Century*, we can see a big difference, especially from the point of view of rewriting the cultural history narrative.

We can see that this kind of rewriting in the digital environment adds three new aspects:

1. deconstruction of the traditional linear image of the past in order to show smaller units, such as events, objects, persons, texts and others;
2. by using these units, representation of the past in a new way, integrating the possibilities offered by special software and/or a hypertext; and
3. integrating the large digital collections made at memory institutions.

Thus, the digitisation processes can be influenced by the needs and requirements of memory institutions, as well as of those of the cultural and literary history writer(s) who use cultural heritage collections to construct cultural and literary historiography in the electronic environment. We can therefore ask the brave question of whether the computational media could become the basis of 21st century historiography and the space for constructing the new architecture of the cultural history narrative. In the digital environment we can visualise and, via semantic relations, display archival sources of cultural heritage (e.g. documents, manuscripts, literary texts) which could play a meaningful role in the creation of literary memories in different periods.

5. Conclusion

Regarding the development of information technology, we could predict, based on the results of the three Estonian applied projects that the new software would enable us to realise much more complicated systems theories in the digital environment, and to develop non-linear space-time models. The constructions of such models basically correspond to the new, non-linear literary history narrative as an open-ended and meaningful labyrinth with different entrances and galleries. These network-like textual models are generated by a principally new type of cultural unit, namely the intertextual thread, which may form new meaningful relations between the macro-elements of literary history. We can compare these remediation processes with translation processes and say that the newest challenge is to 'translate' the linear narrative of cultural history into the non-linear form using the digitised sources from memory institutions.

The possibilities to create such network-like knowledge models, and to make the relations between the historical objects in digital space visible, expand and enrich the notion of the 'user'. Using the heritage collections of the memory institutions, the 'user' can, in addition to being a member of social networks or a partner of the machine, also be the remediator, the architect of the new cultural history narrative. This could be one possibility to overcome the potential miscommunication between the heritage holders in memory institutions and the contemporary "active, versatile users" – a generation *online*. At the same time, we need to remember the fact that the active part of the new media audiences is, despite all possibilities, still small and "the majority lack skills and the will for active participation in the online content" as we pointed out in *Re-Writing Literary Past in the Digital Age* (Laak, Pruulmann-Vengerfeldt 173). Thus, the future of the creative usage of the huge cultural heritage collections laying in memory institution storage facilities could be related to the bridges between the realms of educational and cultural preservation. Only then could skills and active literacy increase and the 'building' of our cultural past in digital space would be full of people who could become the creators of our future.

Acknowledgements

The preparation of this paper was supported by research grants no. 7162 ("Problems of Transformation of Reception of Cultural Heritage in the Digital Age") and no. 8006 ("Developing Museum Communication in the 21st Century Information Environment") financed by the Estonian Science Foundation, and target financed project no. SF0030065s08 ("Sources of Cultural History and Contexts of Literature").

References

Aarseth, Espen J. *Cybertext. Perspectives on Ergodic Literature*. Baltimore, MD; London: The Johns Hopkins University Press, 1997. Print.

Aljas, Agnes, Pille Pruulmann-Vengerfeldt. "Mäluasutuste töötajad digitaalse kultuuripärandi tarbijatena ja loojatena." *Eesti Rahva Muuseumi Aastaraamat* 52 (2009). 114-129. Print.

Bolter, Jay David. "Theory and Practice in New Media Studies." *Digital Media Revisited: Theoretical and Conceptual Innovation in Digital Domains*. Eds. Gunnar Liestøl, Andrew Morrisson and Terje Rasmussen. Cambridge, MA; London: MIT Press, 2003. 15-33. Print.

Even-Zohar, Itamar. "The Polysystem Hypothesis Revisited." *Papers in Historical Poetics. Papers on Poetics and Semiotics* 8. Eds. B. Hrushovski and I. Even-Zohar. Tel Aviv: The Porter Institute for Poetics and Semiotics Tel Aviv University, 1978. 28-35. Web. 17. July 2012.

Jenkins, Henry. *Convergence Culture: Where Old and New Media Collide*. New York; London: New York University Press, 2008. Print.

Jürgenson, Dietrich Heinrich. "Kurze Geschichte der ehstnischen Literatur." *Verhandlungen der gelehrten Estnischen Gesellschaft zu Dorpat* 1. 2-3 (1843-1844): 40-52, 61-73. Print.

Koskimaa, Raine. *Digital Literature. From Text to Hypertext and Beyond.* Jyväskylä, May 2000. Web. 17. July 2012.

Kreutzwald, Friedrich Reinhold. *Kalewipoeg, eine Estnische Sage. Verhandlungen der Gelehrten Estnischen Gesellschaft* 4. 1-4, 5. 1-2/3 (1857-1861). Web. 20 Mar. 2012.

Kress, Gunter. *Literacy in the New Media Age.* London; New York: Routledge, 2004. Print.

—. "Visual and Verbal Models of Representation in Electronically Mediated Communication: the Potentials of New Media Forms of Text." *Page to Screen: Taking Literacy into the Electronic Era.* Ed. Ilana Snyder. London; New York: Routledge, 1998. 53-79. Print.

Laak, Marin. "Beyond Digital Labyrinth: Rethinking Literary History in the Age of New Media." *Different Inputs – Same Output? Autonomy and Dependence of the Arts under Different Social-Economic Conditions: The Estonian Example. Studia Fenno-Ugrica Groningana* 3. Ed. Cornelius Hasselblatt: Maastricht: Shaker, 2006. 48-61. Print.

—. "Eesti kirjandusloo traditsioonist ja võrgukeskkonnast: ERNI kui mudel." *Keel ja Kirjandus*, 2004, nr 10. 721-735. Print.

—. *Kirjandusloo mittelineaarsed mudelid: teksti ja konteksti probleeme* [Non-linear Models of Literary History: The Problems of Text and Context in the Digital Environment]. Doctoral dissertation thesis. University of Tartu, 2006. Tartu: Tartu Ülikooli Kirjastus, 2006. Web. 19 Mar. 2012. http://dspace. utlib.ee/dspace/handle/10062/632.

—. "Cultural Heritage and Literary Memory: Constructing the Estonian Cultural Historcal Web." *Transforming Culture in the Digital Age.* Eds. Agnes Aljas et al. Tartu: Estonian National Museum, Estonian Literary Museum, University of Tartu, 2010. 231-238. Web. dspace.utlib.ee/dspace/bitstream/10062/14768/.../transform_book.pdf.

Laak, Marin, and Pille Pruulmann-Vengerfeldt. "Literary Past as the Cultural Memory: Construction and Reception in the Age of New Media." *Digital Memory: New Media for Preserving the Past. Global Interdisciplinary Research Studies.* Eds. Daniel Riha, and Anna Maj. Amsterdam, New York: Rodophi, 2012. [in print]

— "Re-Writing Literary Past in the Digital Age." *Digital Memories: Exploring Critical Issues.* Eds. Anna Maj and Daniel Riha. Oxford: Inter-Disciplinary Press, 2009. 167-176. Web. 20 Mar. 2012. http://www.inter-disciplinary.net/publishing/id-press/ebooks/digital-memories-exploring-critical-issues.

Laak, Marin, and Piret Viires. "Digital Environment as a Constructor of Literary History: Estonian Experience and Expectations." *eCulture Horizons: From Digitisation to Creating Cultural Experience(s). Salzburg Research Symposium 2004, 27.-28.* September 2004. Salzburg Research Forschungsgesellschaft, 2004. Web. 20. Nov. 2011. http://eculture.salzburgresearch.at/2003/presentation_laak_viires.pdf.

—. "Intertextuality and Technology: The Models of "Kalevipoeg"." *Intertextuality and Intersemiosis.* Eds. Marina Grichakova and Markku Lehtimäki. Tartu: Tartu University Press, 2004. 287-312. Print.

—. "Representing Canonical Texts in the Digital Age: The Case of the Estonian National Epic *Kalevipoeg.*" *Texto Digital* 6.2 (2010): 156-167. Web. 20 Mar. 2012. www.periodicos.ufsc.br/index.php/textodigital/article/view/1807-9288.2010v6n2p156.

Landow, Georg P., ed. *Hyper/Text/Theory*. Baltimore, MD; London: The Johns Hopkins University Press, 1994. Print.

Landow, Georg P., ed. "The Paradigm is More Important Than the Purchase: Educational Innovation and Hypertext Theory." *Digital Media Revisited: Theoretical and Conceptual Innovations in Digital Domains.* Eds. Gunnar Liestøl, Andrew Morrisson and Terje Rasmussen. Cambridge, MA; London: MIT Press, 2003. 35-64. Print.

—. *Hypertext 3.0: Critical Theory and New Media in an Era of Globalization.* Baltimore, MD; London: Johns Hopkins University Press, 2006. Print.

Livingstone, Sonja. *Young People and New Media: Childhood and the Changing Media Environment.* London; Thousand Oaks, CA; New Dehli: Sage Publications, 2003. Print.

Manovich, Lev. *The Aesthetics of Virtual Worlds: Report from Los Angeles.* Nov. 1996. Web. 20 Mar. 2012.

—. *The Language of New Media.* Cambridge, MA; London: MIT Press, 2001. Print.

Pruulmann-Vengerfeldt, Pille, and Agnes Aljas. "Digital Cultural Heritage – Challenging Museums, Archives and Users." *Journal of Ethnology and Folkloristics* 3.1 (2009): 109-127. Print.

Psarra, Sophia. "'The Book and the Labyrinth Were One and the Same.' Narrative Architecture in Borghes' Fictions." *Architecture and Narrative: The Formation of Space and Cultural Meaning.* London; New York: Routledge, 2009. 67-88. Print.

Rojola, Lea. "Uus minevik. Kirjanduse ajalugu uute väljakutsete ees." *Traditsioon & pluralism: kirjanduskonverentside materjale. Ettekandeid, artikleid, esseid ja eksperimendikirjeldusi eesti ja soome uuemast kirjandusteadusest.* Ed. Marin Laak. Tartu-Tallinn: Eesti Kirjandusmuuseum-Tuum, 1998. 256-268. Print.

Saariluoma, Liisa. "Kirjanduslugu eile ja täna." *Akadeemia* 9 (1997): 1885-1914. Print.

Web resources

CULTOS. Multimedia Knowledge Management Tools for Culture and Arts. Salzburg Research, 2003. Web. 20. Mar. 2010. http://www.cultos.org.

DEA. Digiteeritud eesti ajalehed. Valik ajalehti aastatest 1821-1944. Eesti Rahvusraamatukogu, Eesti Kirjandusmuuseumi Arhiivraamatukogu, Tallinna Ülikooli Akadeemiline Raamatukogu, 2003. Web. 20 Mar. 2012. http://dea.nlib.ee.

ERNI. Eesti Kirjanduslugu tekstides 1924-25 [Estonian Literary History in Texts 1924 to 1925]. Estonian Literary Museum, 1997. Web. 20 Mar. 2012. http://www2.kirmus.ee/erni.

Kreutzwaldi sajand. Eesti kultuurilooline veeb [The Century of Kreutzwald. Estonian Cultural History Web]. Estonian Literary Museum, 2010. Web. 20 Mar. 2012. http://kreutzwald. kirmus.ee.

Contemporary Postcolonial and Postimperial Literature in English. Brown University, 22 Mar. 2006. Web. 20 Mar. 2012. http://www.postcolonialweb.org.

The Victorian Web. Brown University. 14 Jan. 2007. Web. 20 Mar. 2012. http://www.victorianweb.org.

Marin Laak, PhD, is is a Senior Researcher at the Estonian Literary Museum, Cultural History Archives. She is the leader of the research project "Sources of Cultural History and Contexts of Literarure" and the web projects on cultural history since 1997. Her current research focuses on the representation models of cultural heritage, specifically in the context of archives and the global processes of computing culture. She is also editor-in-chief of the academic journal *Methis. Studia humaniora Estonica.*

Can You Be Friends with an Art Museum? Rethinking the Art Museum with Facebook

Katrine Damkjær

Communication and Management Employee
Louisiana Museum of Modern Art

Lea Schick

PhD Student
IT University Copenhagen

1. Introduction

The rise of social media marks a paradigm shift in our contemporary culture, to which art museums must respond if they wish to remain relevant in our contemporary society and reach out to new generations of users. This article is built upon an empirical study of the use of Facebook by Danish state-subsidised art museums, showing that it has proven to be a challenge for the museums to integrate the social media into their museum praxis.

Social media have given the broader public the opportunity to upload and share information, and to communicate and interact via online platforms. We see a rising tendency for user participation – users contributing content and no longer satisfied simply being passive consumers. In online social networks the users have become so-called *prosumers*, a term which describes their dual role as both *producers* and *consumers* of content. They are becoming co-creators of our culture, as Dan Tapscott and Anthony Williams describe it. A *prosumer culture* is emerging in which culture is simultaneously produced and consumed by participating users who upload, share, comment on, and produce content.

In this new landscape of a prosumer culture, crucial questions emerge for the art museum as institution – how does this prosumer culture manifest itself in the art museum's praxis? What does this mean to the relationship between users and art museums, and how does it challenge the character and role of the art museum as an institution?

Many art museums are of course already experimenting with social media, creating profiles on social networking services such as Facebook, Twitter, MySpace, and YouTube. In Denmark almost half of the state-subsidised art museums have created a profile on Facebook. This article is based on a recent explorative field study on the use and understanding of Facebook by Danish state-subsidised art museums (Damkjær). The main aim of the study was to document and describe the art museums' use of Facebook during the crucial period in which the Danish art museums made their entrance onto the social media scene. The study consists of an online survey sent out to all Danish state-subsidised museums, investigating their use of social media in general, and their use of Facebook in particular. Furthermore, the study includes qualitative interviews with selected art museums, and detailed observations of the art museums' Facebook profiles, registering the growth of community memberships, level of activity, interaction with users, and types of status updates. The study revealed that the majority of the art museums do not use Facebook on the terms of social media as a platform, failing to make use of Facebook's potential to establish a prosumer culture. Instead, art museums mainly use Facebook as a traditional marketing channel to announce exhibitions and events. They open space for a dialogue neither with nor between users; rather, they communicate *to* instead of *with* users. The study in particular revealed three prevailing dilemmas, occurring when the art museum as institution wishes to expand into social media: 1) establishing a social relationship with users, 2) creating dialogical content, and 3) the convergence of the media. Elaborating on these three dilemmas, we will in this article investigate why and how the art museum as institution must re-think itself in order to fully use social media in its praxis. In the first part of the article, we will outline and elaborate on Damkjær's empirical study in order to discuss these three dilemmas and suggest how they challenge the museum to rethink its traditional character and role in society. In the second part of the article, we will explore how the thoughts of the French philosophers Gilles Deleuze and Félix Guattari can serve as an inspiration and help the museums in this challenging task of both redefining themselves and their relationship with users.

2. The current state of affairs between Danish art museums and Facebook

As one of the most popular social networking sites in Denmark, Facebook has naturally experienced increasing participation by Danish state-subsidised art museums. By September 2009, 18 out of 39 Danish state-subsidised art museums had created a Facebook profile, including three of the largest art museums in Denmark – The National Gallery of Denmark, Arken, and the Louisiana Museum of Modern Art.[1] It is worth noting that in the study, the majority of the larger art

1 The results and examples mentioned above are all from 2009 document and illustrate the

museums are present on Facebook, whereas only half of the smaller museums use the platform. According to the online survey, this is due to a lack of resources (Damkjær). Furthermore, the size of the online community closely mirrors the size and the public display of the physical art museum. The larger and best-known art museums have the largest and fastest-growing communities – the Louisiana Museum of Modern Art has over 10.744 members (22.09.2009) and counting; by comparison, Arken has 1233 members (22.09.2009) and The National Gallery of Denmark has 990 members (22.09.2009). On the other hand, Skagen Kunstmuseum, a medium-sized art museum, has a smaller community with 287 members (22.09.2009), and Kunstmuseet Fuglsang, a small local art museum, has only 56 members (22.09.2009). The outstandingly large online community that Louisiana Museum of Modern Art has managed to attract is greatly in excess of other large museums. According to the study, this is not due to their use of Facebook, which does not differ remarkably from the other art museums. Rather, their large community can be explained by the museum's strong brand and the personal attachment to the museum, which many of their users share (Damkjær). A closer look at interactivity among the users and the art museum in the profiles reveals that the size of the community has little to do with the level of activity. Often the smaller communities of the local art museums have far more activity in comparison to the larger art museums' correspondingly larger communities.

On Facebook, the art museum enters the above described prosumer culture in which users have the urge to be both consumers and producers of the art museums' online profiles by sharing, commenting on, uploading, and tagging content. However, the study of the 18 Danish state-subsidised art museums' use of Facebook revealed that a prosumer culture is barely coming into existence. The content produced by the users is very limited and the engagement in the museum's praxis is restricted. Most of the discussion on Facebook rarely advances beyond small talk and the shared content lacks any immediately apparent art-related, theoretical or cultural importance. Having one of the most active communities among the Danish state-subsidised art museums, Arken receives an average of seven user comments to each of their status updates: *Arken* (03.06.2009)*: "has gone to Venice..."*

User 1:"...to? :)"

User 2: "Surely to take a stroll over the Ponte dei Sospiri, visit the Per San Marco, feel the atmosphere by walking along the canals, look at architecture, art..."

Danish state subsidised art museums entrance and use of Facebook the first five months. It is necessary to make reservations for alterations and new trends in the art museums use of Facebook since September 2009.

User 1: "Eva…in this case I would like to send an application to Arken:)"

Arken: "It is actually because of the Venice Biennale…http://labiennale.org/en"

User 2: Tobias Møller – one of my glass artists participated a couple of years ago. Exiting! I hope you write something or post pictures from the Biennale on FACEBOOK"[2]

In this update, Arken approaches the users directly with a short update informing them about an everyday activity, in the tone and language typically used on Facebook. However, in the update a clash between Arken's identity as an art museum and the personified action, travel to Venice, appears. Can an art museum travel to Venice? The users engage in a humorous conversation wondering what Arken is doing in Venice. Specifically, one of the users asks for descriptions or pictures from the Venice Biennale, but Arken never replies to this request, hereby missing a good opportunity to involve the users in a meaningful dialogue about their work. The value in this conversation is more of a social character, strengthening the social bond between Arken and its users, rather than engaging the users in their experiences and working with contemporary art from the Venice Biennale. The reasons behind the high level of activity in Arken's community is most likely to be found in their friendly and direct approach to users, which generates a continuous everyday-life dialogue with them (Damkjær). However, Arken seems to have difficulties in meeting users on an equal level and engaging them in the museum's praxis. Furthermore, they hesitate to open up to discussions with the users about the museum's exhibitions and work behind the scenes.

2.2 Challenging dialogue

The most common types of comments from users are related to museum visits: users either thank the museum for a pleasant experience or announce their intention of making a future visit to the museum (Damkjær). This type of behaviour is especially common on Louisiana Museum of Modern Art's Facebook profile:

Louisiana Museum of Modern Art (04.09.2009): "Exhibition: Contemporary Art by 24 International artists, Sept. 5th 09 - Jan. 10th. 10"

User 1: "Stunning images ~"

User 2: "Want yours!"

User 3: "longing for you, it is so lovely and nice"

User 4: "just returned home from Louisiana! The deep frozen car in minus 10 degrees is fascinating…"

---

2 All examples are translated from Danish by the authors. The users are anonymised to protect their identities.

User 5: "...lovely:-)"

User 6: "gotta visit!"

User 7: "looking forward to visit Louisiana again next summer"

In this example, the users take part in the marketing of the museum by giving good recommendations and reviews of the museum and its exhibitions. This is the most common role of users in the prosumer culture seen among the Danish state-subsidised art museums on Facebook. A possible explanation for this weak prosumer culture can be found in the fact that art museums mainly use Facebook to market themselves by uploading information about current and forthcoming exhibitions and events. The update above is written in an official language and not directly addressed to users. Louisiana leaves no opening for a dialogue about the exhibition. Facebook is primarily used as a marketing *channel to* the users, and not as an environment within which to *interact with* them. There is a tendency among the majority of the art museums for irregular updates and a generally inconsistent presence on Facebook. Typically the art museum will upload several updates similar to the abovementioned example in one day, then leave the profile inactive for several days or even weeks afterwards. This resembles the use of a familiar medium – the ordinary homepage.

Summing up, these art museums mainly practise monologue rather than dialogue, and fail to encourage users to interact and participate. Even though attempts are made to establish a dialogue with users, as seen with Arken, they do not successfully meet users on an equal level, nor do they engage them in museum praxis. Using the metaphors used on Facebook itself to describe their behaviour, one could say that art museums behave as authoritarian and self-promoting friends.

3. The three dilemmas

Even though increasing interest in user participation is apparent among art museums, integration of Facebook in a museum's practice and the establishment of a prosumer culture seem incompatible with the museum in its traditional form. According to the Canadian Sociologist Carol Duncan, as we know it today the traditional museum institution has strong roots in the Age of Enlightenment, and has mainly had tried to educate the public and present a history – to decide what is included in the narrative of the museum's discipline and what is not, and to shape a national cultural identity. The museum has from its inception possessed an authority and a privileged position as producer of meaning and subject of enunciation, whereas visitors have been passive receivers. Today, this still forms the foundation of the museum and its values, as expressed in the five pillars of museum praxis: to acquire, to register, to conserve, to research, and to communicate, as ICOM defines the museum: "A museum is a non-profit, permanent institution in the service of society and its development, open to the public, which acquires, conserves, researches, communicates and exhibits the tangible and intangible heritage of humanity and its environment for the purposes of education, study and enjoyment" (Museum Definition). Duncan underlines that the art museum, as a social construction, is based on the separation between secular and profane time and space. This separation is revealed in two aspects. First, in the separation between the public space outside the museum walls and the 'sublime' exhibition space within. Second, in the separation of time and space that occurs during the museum visit itself. (Duncan 7) The space of the art museum closes itself around its collection and becomes a space of exclusion, allowing the exposition of the works to reflect, in a controlled context, the values that the art museum wishes to authorise. The museum space is based on its content, and is designed for one-way communication and individual experiences, not for a social purpose. This obviously clashes with the nature of social media, which is built on social relations, changing content, and dialogue, as described by several of the leading theorists within the social media scene – Clay Shirky and Danah Boyd, among others.

The analysis of the art museums' use and understanding of Facebook uncovered three dilemmas in the interrelations between the individual and the institutional as well as between the personal and the public spheres, indicating a gap between a prosumer culture and the traditional art museum:

> The first dilemma occurs when the art museum has to establish a *social relationship* with the users. The dilemma appears in the clash of genres between the art museum as an institution and the art museum as a personal acquaintance. Arkens' update *"has gone to Venice..."* clearly demonstrates this clash of genres. The dilemma is instigated by the fusion of public communication and personal publication, which is present in social media. Furthermore, establishing a social relationship with the users creates difficulty for the art museum, because in this context the institution must be on par with users, and thereby give up the centuries of authority that it has traditionally held. In doing so, the art museum must step down from the hallowed Pantheon of its exclusivity – an act which heralds a breakdown in traditional conceptions of museum space versus public space. The user's personal experience of the art museum becomes part of the museum space, and the museum space expands into public space in the social media.

Secondly, the art museum must first create objects of sociality before meaningful *dialogical content* can manifest itself. Facebook is centred on social networks and relationships, focusing on relationships rather than content. By letting the art museum itself, the art experience, and the museum visit be a social object for interaction, the art museum is forced to open up traditionally closed conceptions of museum space, shattering the one-way communication which museums have mainly practised. In addition, a power shift is emerging in the relationship between users and the art museum, since users through the dialogue take part in the production, presentation and communication of culture. This creates a dilemma for art museums, because they are again forced to release their attachment, as institutions, to the traditional exhibition space and their privileged position.

Thirdly, the art museum has to integrate Facebook into its daily work. This occurrence is a dilemma for the art museum because Facebook represents a convergence of several media forms, mixing communication, branding and marketing. This *media convergence* results in a deconstruction of the art museum's customary structure as an organisation. To handle this media convergence, the individual art museum has to carry out alterations in the museum's resources and routines. Additionally, a new political dimension influences the organisation of the art museum – as a traditionally hierarchical, top-down organisation, the art museum is again sharply challenged by Facebook, which offers a democratic bottom-up structure.

As the findings from Damkjær's previous research show, art museums' presences on Facebook by no means automatically overcome the challenges and provide the benefits of a prosumer culture. Even though the art museums are attempting to establish user involvement and participation, a prosumer culture to a large extent remains unrealised (Damkjær). In other words, it is not possible to be friends with an art museum in its current form. In order to truly embrace a prosumer culture, art museums must transform themselves as institutions into entities capable of dealing with the abovementioned dilemmas by: 1) giving up their authority in order to become social entities, 2) rethinking the museum space to create an environment capable of supporting social dialogue, and last 3) restructuring the organisation of the art museum in order to achieve greater fluidity in its online presence. Addressing these three dilemmas necessitates an acceptance that the museum does not end with its walls, but rather extends into the social space beyond.

3.1 Re-thinking the museum

While the emergence of new media technologies reinforces a transformation of the art museum, this metamorphosis has emerged from within the museum institution itself as well. Often referred to with the term "new museology", a shift from an *exhibition-centred* to a *visitor-centred* museum is occurring. According to Gail Anderson, this shift marks a broadening of focus from primarily the exhibitions themselves to the role of the museum as a whole in serving the public, as well as *how* the museum exhibits art and communicates its background and perceived cultural significance to its visitors: "The collection holdings are no longer viewed as the primary measure of values for a museum; rather, the relevant and effective role of the museum in service to its public has become the core measuring stick" (Anderson 4).

In this context, Facebook can be seen as a decisive catalyst in reinventing the art museum. On Facebook, the focus is on the relationship to, and dialogue with, users. In a prosumer culture, the art museum's authorised role as an institution representing our cultural heritage is challenged, since the art museum's enunciation becomes only one voice among many. By being *co-producers* of the art museum's representation through its online Facebook profile, users themselves become representatives of the art museum. This causes a dramatic power shift between the art museum and its users: "To control a museum means precisely to control the representation of a community and its highest values and truths" (Duncan 8).

Both the museum space and the public space are transformed by the tendency of social media to break down the borders between these spheres. The museum space – or rather the *museum place* – is becoming a nexus of social interaction. In the distinction between space and place lies awareness of the social and cultural aspects of the users' interaction. According to Steve Harrison and Paul Dourish, an online collaborative environment is a notion of place rather than space: "Physically, a place is a space which is invested with understandings of behavioural appropriateness, cultural expectations, and so forth. We are located in 'space', but we act in 'place'." (3) Due to this expansion in the conception of the art museum, museums are encountering a new, emerging user group, which we will refer to as the 'art interested'. This is a growing user group, consisting of people who for one reason or another do not physically visit the museum, but nevertheless have an interest in interacting with the museum and actively follow its activities online (Damkjær).

In the preceding paragraphs, we have shown how an integration of Facebook and a participatory prosumer culture in the museums' identity and praxis is problematic. As we have argued, the reasons behind this can be found in the character of social media, which shatters the very foundation of the art museum's authority.

4. Lines of flight for a social museum

In the reinvented museum, philosophy translates into practice, and practice reflects philosophy, values and mission (Anderson 5).

As described earlier with Anderson's phrase new museology, a rethinking of the art museum is not something entirely new, but is an ongoing process in which philosophy has played a major role. The concepts and thoughts of Deleuze and Guattari have been widely used in rethinking museum praxis.[3] Even though

3 The concept of *the fold* was an important inspiration for Daniel Liebeskind when he designed the Jewish Museum in Berlin. And the department for new media art at The New

Deleuze and Guattari formulated their philosophy in their book *A Thousand Plateaus* long before the appearance of social media, we believe that their thoughts and concepts can be helpful for museums in rethinking their character and their relations to users when integrating Facebook and user participation into their praxis. Deleuze and Guattari criticise the Modern concept of a centred and indivisible subject, which in many ways corresponds to the core values of the traditional art museum – values which we have already demonstrated to be problematic in a prosumer culture. We will argue that using the philosophy of Deleuze and Guattari can therefore be useful in understanding and dealing with these three dilemmas.

As we have earlier touched upon, traditional museum practice is fundamentally grounded in the Enlightenment and in the 'Modern episteme' (Foucault). The Kantian subject as the centre of meaning has dominated the Modern. The individual has been the one and only producer of meaning and subject of enunciation (ibid.). Likewise, museums have heretofore been based on the idea of unity and essence. Museums have had to build a steady and defined identity, an identity that is identical with itself over time, by which we can recognise museums and through which they receive their meaning and reputation (Fehr). Everybody knows what museums such as the Guggenheim, MOMA, and the Louvre stand for.

Based on our study, we will argue that new media and participatory cultures challenge the Modern episteme and the traditional form and function of the museum, as well as the museum's relationship with its users. To understand and to take full advantage of new media, museums have to think outside of the paradigm through which they have defined themselves, and rethink the very identity or character of the museum, in which the users will take a new and much more important part. We will suggest that the thoughts of Deleuze and Guattari can work as tools of inspiration for museums to rethink and transform themselves from a stable identity and *organ*isation to a deterritorialising process of multiplicities and connectedness – a rhizomic body without organs.

4.1 From a centred subject to a myriad of voices

When a museum invites users to take part in the representation and definition of itself, it challenges the character of the museum as *one* unity, *one* subject of enunciation, always identical with itself. Deleuze and Guattari argue that we should dismiss the Modernist idea of the one, stable, indivisible identity, and instead view the subject/body – in this case the museum – from which meaning is being produced and uttered as decentralised processes facilitated by a myriad of multi-

---

Museum in New York City is called Rhizome.org.

plicities. "[T]he question is precisely whether it is necessary to find oneself" (Deleuze, Guattari 156). Meaning is not something that can be linked back to a single unity, but according to Deleuze and Guattari it is rather something ever-changing, being produced in the interaction between a multiplicity of intensities. If we invite the *prosumer* to define the museum, become subject of enunciation, and create dialogical content, we can no longer hold on to the centred subject and the manifestation of a set identity for the museum. This change of character is closely related to the second dilemma, concerning how the museum will have to give up its authority. With social media and a prosumer culture, the museum has to give up its monopoly of enunciation and accept that it is no longer the sole creator of meaning. There will no longer be one single hierarchical root through which one can trace back the identity of the museum as subject, but instead we must accept that the character and the role of the museum is constantly emerging and evolving through a continuous stream of interactions and participations – within the social media, within the museum space itself, and in other spheres as well. The division between the museum as institution and the prosumers as individuals is eliminated, and instead an interdependent relationship built on participation and interaction is arising. These new users interpret and participate in the making of the museum.

4.2 The museum as a body without organs

Looking back at the third dilemma – illustrating how an integrated use of Facebook causes a deconstruction of the organisation, and of the top-down structure of the museum – we see that the museum's loss of its monopoly as subject of enunciation and producer of meaning causes a breakdown in the functioning of the museum's traditional organs. We can no longer think of communication, branding, curating, etc., as defined and separated authorities because users become an integral part of these functions. Deleuze and Guattari are primarily fighting three concepts strongly linked to the Modern episteme: the centred subject, fixed meaning, and organisms. They suggest that we should instead strive for a *Body without Organs* (BwO) in order to make a break with the Modern way of thinking (Deleuze, Guattari 149-166). BwO is not a battle against the organs as such, but is a fight against the *organ*isation. When we organise – be it museums, politics, our home, or any kind of meaning – we give each entity one, and only one, function. It is clear where a message is coming from, and who decides what. This corresponds very well to the traditional five-pillar organisation of the museum. However, the problem is, Deleuze and Guattari argue, that by organising structure, we leave out the potential for other functions and the possibility for meanings to arise from other kinds of connections, and from the relations between varieties of things and actions. BwO is, say Deleuze and Guattari, to open up the body – in our case the body of the museum – to connections, montages, flows, relations, and possible *lines of flight*. In a BwO there is no centre from where actions are controlled, no single subject of enunciation, no essence, but meaning and functions are being defined in flows between many organs. "It is a question of making a body without organs upon which intensities pass, self and other – not in the name of a higher level of generality or a broader extension, but by virtue of singularities that can no longer be said to be personal, and intensities that can no longer be said to be extensive" (Deleuze, Guattari 156).

If we apply this experiment of thought to the museum and its use of social media, the implication is that the museum should be open to the multifaceted meaning and functions that are being constantly produced and reformulated in a continuous and ever-changing process of interaction between museum, users, and media. Here we cannot separate the museum from its users, "Thus the BwO is never yours or mine. It is always a body." (Ibid. 164) As described earlier, the museum expands beyond its walls, and the users and their interactions – both the interactions online and those onsite become part of the museum place. Even the so-called 'art-interested' who never visit the museum on-site, but only participate from their living rooms become part of the museum's body without organs. This BwO museum is an unorganised place, where the relations between user and museum are constantly forming new constellations. This all leads to a transformation of the museum from a stable entity to a dynamic rhizome.

4.3 Museum as rhizome

In the description of the museum as a place of process and interactions, another of Deleuze and Guattari's philosophical concepts proves useful. In the introduction to their book *A Thousand Plateaus*, they describe one of their most famous ideas: the shift from a hierarchical, linear, and genealogical tree structure to a decentralised, deterritorialising, and antigenealogical root structure – a *rhizome* (Deleuze and Guattari 3-25). The use of the rhizome as a metaphor derives from a specific kind of root structure that doesn't (as *does* for example the root structure of the tree) have one main root to which you can trace any other part of the root. The rhizome is a web of roots that have no centre and no single direction. In a rhizome, every point is connected to every other point, and the connections become more important than the individual points. This matches very well our earlier description of how connections in social media become more important than content. The rhizome is made only of lines: "lines of segmentarity and stratification as its dimensions, and the line of flight or deterritorialization as the maximum dimension after which the multiplicity undergoes metamorphosis, changes in nature" (ibid. 21). Any modification in any connection in this web affects the rest of the rhizome, but damage to any one point does not necessarily lead to wholesale destruction of the network and its components, rather it gives rise to a deterritorialisation. In our case, the museum's loss of monopolistic enunciation and the dissolution of the museum walls do not result in a weakening of the museum but rather facilitate a development and a reconfiguration of the museum.

The museum has typically been built upon the traditional model of the hierarchical tree root structure; in fact, the same structure underlying the thoughts of the Enlightenment – concepts which have, as explained earlier, been the very basis for the museum's role in society. Most museums are built on a genealogical foundation and are organised in a linear or a categorical fashion. The museum is traditionally hierarchically organised and the meaning interpreted from its content can be traced back to the curator and/or to the artist. The museum has hitherto ended at its walls, which have in turn defined the museum space. The rhizome is on the contrary not a delimited space, but rather consists of "multiple entryways and [...] lines of flight" (ibid. 21). These entryways can be the user on site, or the ways we enter the museum through different media. "The rhizome has no beginning or end; it is always in the middle, between things, interbeing, intermezzo" (ibid. 25). When the museum expands beyond its walls and folds out into the social media it too has no end and no centre, but the users become part of the museum, producers of the museum. This implies that the museum must give up its defined and stable identity; instead of *being* something, it must accept that it constantly *becomes* something new in its interaction with the users.

As we have described here, when a museum opens up, expands into social media, and invites the users to be co-creators of the museum's identity, we see that the museum is compelled to think of itself more through the concept of a rhizome, and more as an open space of interactions, than as a defined, rooted, and limited space. Because museums are traditionally accustomed to possessing a relatively fixed identity and hierarchical root-structure, providing the users with the power to co-define the museum can seem dangerous, since the multiplicity of enunciation can change the foundation of the museum and blur its set identity. However, if the museum embraces the concept of the rhizome, it would no longer be a daring process to invite the users. The users would be exactly what constitutes and develops the museum's character, making sure it is changing *with* and thereby staying relevant *to* users.

Using social media, museums have to open up and perceive the comments and interactions of users as lines of flight, to which they should respond and from which they should develop their practice. Equally, museums must think about their content on Facebook as lines of flight – something that can be continued and developed by users. Obviously, this is not done by posting simple, practical information about the exhibitions. The museum has to accept that their challenge is to facilitate lines of flight, that they are simply engines from which cultural production can begin, although the actual creation happens in interaction with prosumers. The users and the media become part of the museum, create lines of flight, and their connections to each other change and redefine the museum. Users, media, museum employees, the walls, the artefacts, the communication, and the visitors are all part of the same rhizome that we here refer to as "the museum".

References

Anderson, Gail, ed. *Reinventing the Museum: Historical and Contemporary Perspectives on the Paradigm Shift.* Walnut Creek, CA; Oxford: Rowman Altamira. 2004. Print.

Boyd, Danah. "The Term Social Software." *Corante: Many-to-Many.* A Group Weblog on Social Software. 14 Oct. 2004. Web. http://many.corante.com/archives/2004/10/14/the_term_social_software.php. 2004.

Damkjær, Katrine. "Overvejer, om man kan være venner med et kunstmuseum...? Kunstmuseets dilemmaer på Facebook." Master thesis for IT-University, Copenhagen. 2009. Unpublished. http://www.smk.dk/fileadmin/user_upload/Billeder/om-museet/museets-projekter/SMK_digital/Speciale_venner%20med%20et%20kunstmuseum.pdf.

Deleuze, Gilles and Guattari, Félix. *A Thousand Plateaus.*1987. Minneapolis, MN; London: University of Minnesota Press. 12th ed. 2007. Print.

Duncan, Carol. *Civilizing Rituals: Inside Public Art Museums.* London: Routledge. 1995. Print.

Fehr, Michael. "Understanding Museums – Das Museum als autopoietisches System." Eds. Michael Fehr, Clemens Krümmel, and Markus Müller. *Platons Höhle. Das Museum und die elektronischen Medien.* Köln: Wienand Medien, 1995. 11-20. Print.

Fehr, Michael. *Text and Context: Developing a Museum by Reflecting its History.* Web. 1. Feb. 2010. www.aesthetischepraxis.de/Texte_English/Fehr_Text&Context.pdf.

Foucault, Michel. *The Order of Things: An Archaeology of the Human Sciences.* 2nd ed. New York: Routledge. 2001. Print.

Harrison, Steve, and Paul Dourish. *Re-Plac-ing Space: The Roles of Place in Space in Collaborative Systems.* Cambridge: Draft submission for CSCW'96, Xerox Palo Alto Research Center & Rank Xerox Research Center, Cambridge Lab. 1996. 67-76. Web. 6 Dec 2010. www.dourish.com/publications/1996/cscw96-place.pdf.

"Museum Definition." ICOM. International Council of Museums. Web. 6 Dec. 2010.

Shirky, Clay. *Here Comes Everybody: The Power of Organizing without Organizing*. New York: The Penguin Press. 2008. Print.

Tapscott, Dan, and Anthony D. Williams. *Wikinomics: How Mass Collaboration Changes Everything*. New York: Portfolio. 2006. Print.

Katrine Damkjær is Communication and Management Employee at Louisiana – Museum of Modern Art. She holds a master in 'Digital Design and Communication' from the IT University of Copenhagen and a Bachelors in Art History from Copenhagen University. http://www.louisiana.dk/dk

Lea Schick, PhD Student, Technologies in Practice faculty group, IT University of Copenhagen. She holds a master in 'Digital Design and Communication' also from the IT University and a Bachelors in 'Aesthetic and Culture' from Århus University. In her PhD project she explores how a new electricity infrastructure based on renewable energy is imagined and made in Denmark. http://itu.dk/tip/?p=1181

Recommended Friends, Artists, Events and Books: The Opportunities and Risks of Web 3.0

Stijn Bannier

Chris Vleugels

IBBT-SMIT, VUB
Studies on Media Information and Telecommunication
Interdisciplinary Institute for Broadband Technology
Vrije Universiteit Brussel
Pleinlaan 9, B-1050 Brussels
Belgium
Bannier@merit.unu.edu
Chris.Vleugels@vub.ac.be

1. Introduction

Guided by the perspectives of Web 2.0 and Web 3.0, this article examines to what extent Web 3.0, and in particular, recommendation engines, will deliver user-generated, machine-mediated content, and what implications this has for cultural participants and their engagement to and with cultural heritage. Therefore, we use the findings from the Flemish IBBT (Interdisciplinary Institute for Broadband Technology) research project CUPID (Cultural Profile and Information Database). This research project searched for innovative ways to aggregate, categorise, personalise and distribute cultural content in order to give end users a rich cultural experience. The outcomes of the studies on the key concepts of Web 2.0 and Web 3.0 will be used to discuss the shift from user-generated to machine-generated content, and the case of the recommendation platform of cultural events and cultural heritage, as developed in the CUPID research project. We use the findings from different series of focus group conversations to reflect on the opportunities and risks of Web 3.0 and personalised recommendations and show how these outcomes will be used in CUPID's follow-up project ICIS (Infrastructure and Culture Information System).

2. From Web 1.0 to Web 2.0 to Web 3.0

The term Web 2.0 is used to describe the evolution (Knowles 3; Skiba et al. 10) from a 'read-only' World Wide Web, later called Web 1.0, towards web sites and web services based on participation, collaboration and sharing between users

(O'Reilly 6; Rollett et al. 4-5). These concepts transcend the former dichotomy of creators and consumers of content. Content is now based on the harnessing of data and outputs of users. Web 2.0 applications allow users and websites to consume and remix data from multiple sources. These applications and websites emphasise interaction, community and openness (Millard and Ross 1). Examples include applications such as weblogs and wikis, technologies such as social tagging and social networking and websites such as YouTube, MySpace and Flickr.

Inventor of the World Wide Web Tim Berners-Lee saw the web not only as a common information source, but also as a place to "work and play and socialize" and where computers can "help us analyse it, make sense of what we are doing, where we individually fit in, and how we can better work together" (Berners-Lee 1). With coining the term Web 2.0, the future of the World Wide Web is also widely discussed. The World Wide Web seems to evolve further. To what extent can a Web 3.0 arise and what does it signify?

To analyse the web and make sense of what users are doing, Berners-Lee et al. see a semantic web as the future of the web, "creating an environment where software agents roaming from page to page can readily carry out sophisticated tasks for users". (3)

The semantic environment will make use of data and metadata, i.e. data on data, the semantics of information and services on the web. The web will become more intelligent and will be able to satisfy and fulfil the requests of users. Adding content and metadata to Web 2.0-applications and organising and processing these data in a Web 3.0 application should make the Internet easier to use (Borland 65).

Although Web 3.0 is seen as the symbiosis of web technologies and knowledge, some theorists are more critical. Among others Lassila and Hendler:

> Although many aspects of the Semantic Web are yet to be explored, and much research remains to be done, this technology is clearly transitioning into a serious player in the modern Web universe. We might not like the term "Web 3.0," but we enthusiastically embrace the technologies it is bringing to the field. (93)

Mike Evans foresees several practical problems with Web 3.0 too (58). Firstly, the semantic web becomes more and more complex and machines cannot understand everything that users put in. Secondly, the democratised production of content by users is being ignored by processing data and generating metadata and thus content in Web 3.0-applications. And finally, Evans indicates that a forced pursuit of the semantic web prevents the dynamics of the World Wide Web. He thinks the Internet will evolve into a more hybrid web – he already speaks of Web 4.0 –, provided by mobile technologies and using physical objects. This is something Berners-Lee et al. also forecasted, when "the Semantic Web will break out of the virtual realm and extend into our physical world" (15). This means a hybrid web, spun from a number of technological threads, all helping to make data more accessible and more useful (Borland 1) and a web about recommendation and personalisation (Kiss 1).

2.1 The user as producer

Not only is the concept of the World Wide Web challenged by the evolution mentioned above, but so too is the role of the user. The demarcations between consumer/producer and amateur/professional seem to blur and create hybrid roles because of new media and digital culture. New media and digital technologies have collapsed the traditional value chain of production, distribution and consumption. Users are now able to produce and distribute their own content on the Internet. Axel Bruns coined the term 'produsage' to describe this "new hybrid form of simultaneous production and usage" (Bruns 99). Since the Internet earlier only showed pre-produced websites (Web 1.0), today the Internet gives lots of possibilities to users to create their own content online (Web 2.0).

The 'produser' transcends the former chasm between user and producer. With active 'produsers' there is more and more involved consumption. In addition, users are trying to produce more professional content as well. This new hybrid role goes by the name 'produser' and was coined alongside 'prosumer' (Toffler 266). Thus, the borders between amateur and professional content are also under pressure, since new media easily facilitate amateurs or consumers to produce professional content themselves, since the "PC itself is in many ways the ultimate figure of media 'prosumer' technology. It is a technology of distribution, of consumption, as well as a technology of production." (Lister et al. 34)

New media and digital technologies facilitate 'produsers' and 'prosumers' to create their own content. Web 2.0 therefore enables users to create and distribute their creations, and consume others' user-generated content.

2.2 From user-generated to machine-generated content

As already stated, Web 3.0 will make data and content more accessible and usable. This offers new possibilities for the organisation of content. For example, Web 2.0 websites like Facebook or Last.Fm are able to include recommendations based on available content. The huge amount of user-generated content of Web 2.0 will be distributed even better because of Web 3.0 filtering techniques of recommending data. The user, although not completely in control of content creation, does well by this development, as the data and content will be recommended and personalised and this is more accessible and useful. However, to what extent the user keeps the same role in the emerging Web 3.0 technologies remains to be seen.

In particular, the user seems to take advantage of Web 3.0, yet the user is no longer the only one adding value to the web. Complex algorithms process data, generate metadata and add content to the Web. Combined with the content obtained via Web 2.0 applications, the connection between large amounts of structured information, and user and context related indicators, will lead to the enrichment of the information. Enrichment refers to relations between similar information (such as text, video and audio), facilitating content (such as public transport, parking facilities, restaurants and weather forecasts) and user-generated content, but also relations between data, metadata, user profiles and contexts. Web 3.0 applications will combine Web 2.0 concepts, which let users generate content and create and update appropriate profiles, and Web 3.0 concepts to create personalised and enriched content for the user.

In real terms, this means that the user will create and update a personal profile with data such as personal interests, artistic preferences and visited museums, expositions and cultural events. Through hybrid filtering algorithms, combining content-based and social-based (collaborative) filtering, metadata will be created for the content. Moreover, the application recommends certain events, museums, expositions, painters, books, etc. This "user-generated, machine-mediated content" (Dahlen) or machine-generated content will be the data of Web 3.0, hence the data that will make other data and content more accessible and usable.

3. CUPID – Cultural Profile and Information Database

CUPID's research focused on innovative ways to aggregate, categorise, personalise and distribute cultural content in order to give end users a rich cultural experience. The core research questions within the project focussed on four related aspects, taking into consideration the needs of both professional users (such as cultural institutions) and end users. Firstly, starting from the supply side, research concentrated on ways to optimise the aggregation and management of cultural and leisure information. Secondly, solutions to categorise cultural and leisure information were worked out. Thirdly, at the demand side of the story, research activities were focused on an appropriate distribution and presentation of content and information for specific target groups. For this purpose an online profiling and recommendation system was developed. Fourthly, alternative distribution models and mechanisms, such as digital television and mobile phone, were investigated. The project was also challenging with regard to legal aspects such as copyright, privacy laws and liability risks for the managers of enriched content.

3.1 CUPID models

In the CUPID project, two models were developed: on one hand, a short-term model, an operational prototype in which the distribution of personalised event information is limited to Facebook; on the other hand, a long-term model (a so-called 'CultureFeed') which will provide the user with detailed possibilities to manage, control and share the information about his cultural participation and which will operate independently of any web platform. Both models start from user-generated profile information, combined with user-generated event information, used to machine generate recommendations for cultural events.

3.1.1 Short-term model

The most important result of the project is a Facebook application as an operational prototype of CUPID, connecting the cultural event database 'UiTdatabank', a recommendation system, with the community of arts centre Vooruit (see figure 1).

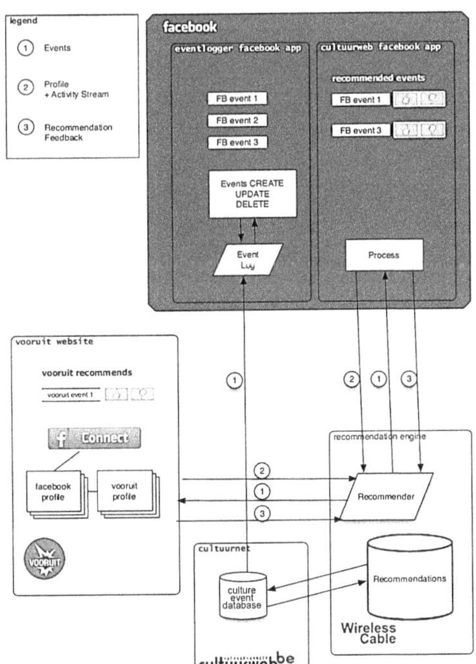

Figure 1: CUPID's short-term model.

The ‹UiTdatabank› is the central database where all the information on leisure and cultural events in Flanders is assembled. In the prototype, all cultural events are transferred from the database to Facebook events. As a result, cultural organisations do not have to create their events on social networking sites themselves, but just have to keep one central database up to date. Next, a public Facebook application provides the user with recommendations of future events on which participation and appreciation feedback can be given. The feedback of the user on the events is sent to the recommendation engine and will be taken into account for future recommendations. As a result of this feedback process, the user builds his or her own cultural profile. When a critical mass of users of the Facebook application is obtained, the recommendation engine can generate even better recommendations on the basis of collaborative filtering. In addition, users who have a profile on the website of the Vooruit arts centre are able to link their 'Vooruit profile' with their profile on Facebook by means of Facebook connect. In consequence the user's ‹Vooruit profile› is enriched by the Facebook profile, and results in better recommendations from events taking place in the arts centre.

3.1.2 Long-term model

The long-term model (see figure 2), the 'CultureFeed', has two main objectives. It will provide the user with more possibilities to manage, control and share information about his or her cultural participation, and it will operate independently of any platform. The user will be able to assemble a personalised calendar of cultural events and get recommendations on the basis of his personal calendar and other cultural information he wants to share. Again, he will have the possibility to give feedback on the recommended events. Next, the user will be able to manage his cultural profile in much more detail. He or she will be able to decide which personal information is private and which information will be shared with the recommendation engine and specific cultural organisations. The user will also have the possibility to add and delete manually specific items of information. Further, the user will have the possibility to log in to the websites of cultural organisations by means of OpenID in order to receive recommendations on these websites. Finally, the user is able to build a social network of 'cultural friends', which he or she can follow and appoint as preferred friends so that they have more weight in the recommendation system based on collaborative filtering.

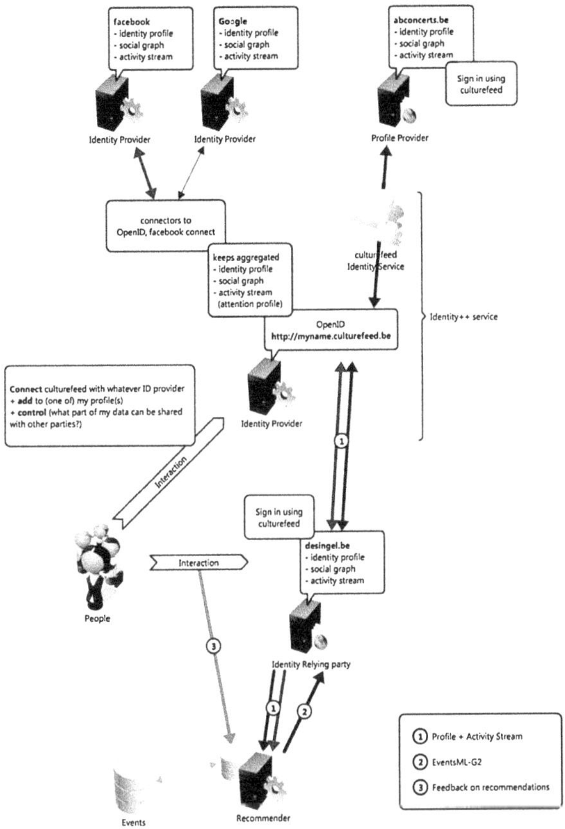

Figure 2: CUPID's long-term model.

3.2 Focus group findings

During the CUPID research process, different series of focus group conversations were carried out in order to reflect on the possible opportunities and risks of combining user-generated machine-generated content in relation to the CUPID models. Focus groups were set up with external experts (technological experts and conceptual experts), cultural mediators (umbrella organisations, cultural organisations, media, cities and municipalities, and social networks and weblogs) and end users to reflect on and evaluate the developed CUPID prototype and its long-term model.

3.2.1 *Experiencing abundance and an emerging demand for recommendations*

Abundance is an argument that was used often during the focus group conversations. People experience an abundance of existing social networking sites, of user-generated content, of invitations by friends or cultural organisations and of online and offline information about cultural events. According to our respondents, this has two major consequences. First, due to an abundance of all these kinds of information, people tend to participate less than they really want to. Second, people tend to fall back on events they already know, with the result that they experience new artists, events or cultural genres less, or not at all.

The event feature on social networking sites is mainly used for parties among friends or for the events of local organisations. For bigger or national cultural events, the calendar tool on social networking sites is hardly used. The event function is therefore dominated by the viral effect of friends sending invitations. Only a few participants indicate that they use social networks to search for interesting cultural events. The advantages of keeping a calendar on a social networking site are twofold. In the first place, it is useful to have a visual overview of events one will or might attend, meaning that the calendar serves as a reminder tool. Secondly, a calendar is useful to see who, from the social network, will go to an event and what comments they have posted. This can have both a positive and a negative influence on their own participation. Following on from this, the privacy argument is used too to justify not keeping a calendar, since some people do not want their whole social network to know where they are going. In addition, by not using the event feature on social networks, some participants indicate they want to prevent an abundance of invitations. When we turn to the use of the social network feature on cultural profile sites such as Last.fm, we note that this feature is less used than the event function on social networking sites. The main argument is that there is already an abundance of social networks and one wants to limit the number of networks on which one is active. In the opinion of our respondents, the purpose of cultural profiles is to discover unknown artists and new cultural genres, rather than looking for members who are also in other social networks, such as Facebook. If one becomes friends on cultural profile websites, it is mainly for practical reasons, for example in order to travel together to events by using a carpool system.

Web 3.0 applications, such as the CUPID system, tend to provide an answer to the problem of abundance by creating personalised and enriched content for the user. Some respondents already experienced recommendation systems, as provided by Amazon, Last.fm or iTunes. The recommendations have led to additional purchases and the discovery of new authors and artists. Some others refer to national cultural organisations who have their own community and who generate recommendations themselves. However, these recommendations are tied to the supply of a particular cultural organisation and respondents prefer a recommendation system that covers the complete cultural sector.

3.2.2 The influence of recommendation on cultural participation and consumption

During the focus groups, the conversation partners gave very positive reactions towards the CUPID models. They experienced getting recommendations, being surprised with unexpected cultural events and making use of an automated data flow as an added value. Important elements of their existing online social networks are now combined with recommendations in a cultural context. Next, new opportunities of CUPID were discussed in the form of the option of saving attended events into an archive and the implementation of user-generated, together with the existing machine-generated, context information.

With regard to the recommendations, the potential end users attached great importance to setting filters in a good and easy way to include the amount of recommendations, including on which days one wants to get recommendations and to include locations and genres in order to get good recommendations. These filters are important to prevent an abundance of recommendations. The focus groups emphasised not only new opportunities to provide additional background information, but also the possibility of viral recommendations (from person to person). In addition to this, they experienced the convenience of implicit user-generated feedback positively (like Last.fm), because updating a profile in this way will be less time consuming.

Although the CUPID project assumed more cultural participation among more people, most focus group participants expect more and better participation among the current cultural participants, because the automated data flow ensures CUPID users to get the information in time, and the recommendations create the possibility to discover new cultural disciplines and events. For their part, cultural organisations valued being part of a recommendation engine as an extra tool in their relationship with their audience. In particular, they valued the automated distribution of their cultural events on several online platforms. Since the recommendation system uses social networks, organisations could be able to reach a larger audience.

3.2.3 Being in the driver's seat

Although end users experience recommendations and surprise at unexpected cultural events as added values, they also indicate privacy as a bottleneck affecting both the degree of openness when sharing personal cultural information, and willingness to share the received recommendations. Concerning the privacy aspects of

CUPID, all participants agree that the user must be able to control all the settings. The user should define what is private or public in the cultural profile. When both profile integration possibilities, and privacy aspects are highly customisable, the participants see no problem in investing a bit more time in a new cultural profile, provided that the CUPID system will work well and the recommendations meet their expectations. Furthermore, a lack of experience with OpenID emerged from the focus group conversations. Although most participants know the concept of OpenID and consider using one login and password as convenient, most of them have little or no experience with this standard or have knowledge of issues that possible damage privacy.

Web 3.0 might be semantic and intelligent, and data and content might be more usable and better accessible, but machine-generated content is also less controllable than Web 2.0's user-generated content. To what extent an abundance of user-generated content is replaced by an abundance of machine-generated content is something that should be taken into consideration. Will the overload of Web 2.0 profile information not just result in an overload of recommendations in Web 3.0? The CUPID research showed that a certain fear for abundance of recommendations exists amongst the participants. Such an overload can be the result of very detailed or diverse cultural profiles. In addition, user-generated content does not always reflect the cultural interests of a person because some content may not be applicable to the profile (for example "just for fun" groups or certain fan pages on Facebook), but is nevertheless used to generate recommendations. In consequence, an abundance of recommendations might occur. This could result in users no longer looking at personalised and enriched content. In this case, Web 3.0 applications would be back to square one, since the user is still confronted with an overload of user- and machine-generated content. In addition, some participants mentioned a possible impoverishment of the cultural experience when users are guided only, or overly, by recommendations. As a result, when relying completely on machine-generated recommendations, users could miss interesting cultural events, or could become more indolent regarding searching for interesting cultural events.

4. ICIS: Infrastructure and Culture Information System

With CUPID's short- and long-term models in mind, the research project's outcomes were used to create a follow-up project, called ICIS (Infrastructure & Culture Information System). This public information service will aggregate all available data on culture, heritage and leisure and combine it with personal profiles and location-based applications. By linking information on cultural heritage and extensive leisure agenda ICIS opens many new possibilities. Everyone is able to create their own trail through the cultural landscape in Flanders and become more aware of cultural and historical surroundings. The aim of the ICIS is to create a way to make cultural and historical heritage optimally accessible.

Currently, the ICIS project is in its start-up phase. The project will include a mobile application, a website and possible interactive stations to log a personal profile, enter one's wishes and create a possible cultural or leisure program. Thus, a user could log in on his or her mobile ICIS application using a personal cultural profile for cultural and leisure history. Then, leisure duration and other aspects can be specified, such as distance or location. Finally, ICIS will create a personal recommendation in the form of a possible museum or concert visit, a cultural heritage itinerary or any other cultural or leisure event or location. By linking people's personal interests, geolocation data, information on cultural heritage and an extensive leisure agenda, ICIS wants to realise CUPID's goals by forming a digital tool to facilitate a user-centred approach in the cultural field.

5. Conclusion

This article discussed the changing role of the user in view of the cultural transformation from user to producer facilitated by Web 2.0, and what changes might be expected from Web 3.0 and its machine-generated data. Web 2.0, the current phase of the World Wide Web and online applications, is characterised by interactive and dynamic content. Web 3.0 is assumed to comprise the following opportunities: a hybrid, semantic and intelligent web made possible by the convergence of several new technologies, which will make data and content more usable and better accessible. The CUPID research project searched for innovative ways to aggregate, categorise, personalise and distribute cultural content in order to give end users a rich cultural experience. As a result, the CUPID short-term prototype as well as the long-term model combines concepts of Web 2.0 and 3.0.

When the World Wide Web changed from 1.0 to 2.0 to 3.0, the user changed respectively from a viewer to a produser/prosumer, to a producer and a user of personalised content. This personalised content is both user-generated and machine-generated. On the one hand this offers lots of new possibilities to the user, since it makes content better accessible and usable. On the other hand, it scares users, since they are no longer completely in control with regard to privacy options, abundance of recommendations or missing certain information because of a (machine) predefined selection of content. These latter aspects require more research, since transparency of how recommendations are generated is required. Moreover, further information regarding the support of privacy issues should be provided. In addition to this, the CUPID research project focused mainly on a short-term model, an operational prototype. The long-term model should include more possibilities to manage, control and share information about cultural participation and operate platform-independently. A long-term model should also be able to give valuable and useful intercultural recommendations.

CUPID's follow up ICIS combines the concept of the changing user in light of the Web 1.0 - 2.0 - 3.0 evolution, the institutional and individual cultural transformation processes because of digitisation and the newest technological developments, such as personalised recommendations and geolocation tools. This combination of user-generated and machine-generated content not only results in a complex interaction and convergence between all types of cultural actors (producers, distributors, institutions and users), but will also result in reciprocal crossovers and self-reflection about the roles and statuses of different actors in an online and offline context.

References

Berners-Lee, Tim, James Hendler, and Ora Lassila. "The Semantic Web – A New Form of Web Content that Is Meaningful to Computers Will Unleash a Revolution of New Possibilities." *Scientific American* 284.5. (2001): 34-43. Print.

Berners-Lee, Tim. *The World Wide Web: A Very Short Personal History*. 7 May 1998. Web. 2 Mar. 2010. http://www.w3.org/People/Berners-Lee/ShortHistory.

Borland, John. "A Smarter Web." *Technology Review* 110.2. (2007): 64-71. Print.

Bruns, Axel. "Produsage: Towards a Broader Framework for User-Led Content Creation." *Creativity & Cognition 2007: Seeding Creativity: Tools, Media, and Environments (Washington, DC, 13-15 June 2007)*. Eds. Gerhard Fischer et al. New York: Association for Computing Machinery, 2007. 99-106. Print.

Dahlen, Chris. *User-Generated, Machine-Mediated Content*. 26 Dec. 2008. Web. 2 Mar. 2010. http://savetherobot.wordpress.com/2008/12/26/user-generated-machine-mediated-content/.

Evans, Mike. *The Evolution of the Web – From Web 1.0 to Web 4.0*. Reading: University of Reading, 2007. Lecture.

Kiss, Jemima. "Web 3.0 Is All about Rank and Recommendation." *The Guardian*. 4 Feb. 2008. Print.

Knowles, Julian. "A Survey of Web 2.0 Music Trends and Some Implications for Tertiary Music Communities." *Music in Australian Tertiary Institutions: Issues for the 21st Century, June 29-July 1*. Ed. Catherine Grant. Brisbane: Queensland Conservatorium Griffith University, 2007. Print.

Lassila, Ora, and James Hendler. "Embracing "Web 3.0"." *IEEE Internet Computing*. 11.3. (2007): 90-93. Print.

Lister, Martin. et al. *New Media: A Critical Introduction*. London: Routledge. 2003. Print.

Millard, David E., and Martin Ross. "Web 2.0: Hypertext by Any Other Name?." *HYPERTEXT '06: Proceedings of the Seventeenth Conference on Hypertext and Hypermedia*. New York: ACM Press. 2006. 27-30. Print.

O'Reilly, Tim. *What Is Web 2.0? Design Patterns and Business Models for the Next Generation of Software.* 30 Sep. 2005. Web. 2 Mar. 2010. http://oreilly.com/web2/archive/what-is-web-20.html.

Rollett, Herwig et al. "The Web 2.0 Way of Learning with Technologies." *International Journal of Learning Technology* 3.1. (2007): 87-107. Print.

Skiba, Brian, Alexander Tamas, and Keith Robinson. *Web 2.0: Hype or Reality... and How Will it Play out? A Strategic Analysis.* Arma Partners. 9 Feb. 2006. Web. 2 Mar. 2010. http://www.armapartners.com/files/admin/uploads/W17_F_1873_8699.pdf.

Toffler, Alvin. *The Third Wave.* New York: Bantam Books, 1980. Print.

Stijn Bannier is MA in New Media and Digital Culture and worked from October 2008 until December 2011 as an e-culture researcher at IBBT-SMIT, Vrije Universiteit Brussel. In January 2011, Stijn joined the Collaborative Creativity Group at UNU-MERIT, Maastricht University, working on research projects on intercreativity, digital music culture and digital publishing.

Chris Vleugel is a Sociologist who worked from October 2006 until May 2012 as a researcher at IBBT-SMIT, Vrije Universiteit Brussel on interdisciplinary projects regarding new technologies within the cultural field. Since May 2012 he is part of The Flemish Office of the Children's Commissioner where he is responsible for the fields of education, leisure and media.

Blogging Writers: (De)Mystification of Authority?

Joke Beyl

PhD Student
iMinds-SMIT, Vrije Universiteit Brussel
Pleinlaan 2
1050 Brussels
Belgium
joke.beyl@vub.ac.be

Introduction

In this article we start from the idea of the weblog as a communicative and interactive medium. Taking this notion into account we then question, firstly, what this medium means in terms of representing oneself as a literary writer; secondly, we try to understand how traditional communicative relations are reflected upon in this particular context. Seeing that digital interactivity through weblogs both allows conventional cultural content producers to express a self-constructed image (boyd) and, at the same time, allocates a more active and creative role to the audience (Cammaerts 358), our study concentrates on unveiling the online narratives that blogging writers construct about themselves, about their role and position in society and about their relationship with the audience. This article, then, first intends to theoretically explore how weblogs might influence writers' self-conceptualisation, building on a historical study of the conceptualisation of artisthood as well as on several contemporary reflections about media and interactivity. Second, we empirically elaborate on these theoretical suggestions by means of an extensive case study of a specific literary writer's weblog. This is done through a narrative analysis of the writer's blog posts that were published in January 2008 to find out to what extent the writer uses the weblog as a medium to (de)mystify his aesthetic role and authority position in society. In addition, we expand this analysis with insights obtained from an in-depth interview with the writer in question to understand how he perceives these blogging activities. In sum, this article – which is related to the author's PhD-research dealing with multiple case studies, one of which will be discussed in this article – seeks to offer both a theoretical framework and a concrete empirical elaboration on the way the writer's weblog, as a self-presentational and interactive medium, can shape the image of the literary author.

Writers and authority

16th century Europe experienced a conceptual shift from the artist as mere artisan to the artist admired for his innate creativity, ability and exceptionality (Barker et al. 7). However, it was not until the end of the 18th century, during Romanticism, that the notion of the artist as genius became acknowledged on a large scale. This notion referred to the artist as a creative and autonomous individual deciding independently about a work of art. Gaining this type of autonomy meant that the artist no longer merely created in compliance with the wishes of others (Van den Braembussche 193-194). This also initiated a split between what was regarded as science, i.e. a communal and objective description of the world, and art as a solitary and subjective experience. For the artist now became depicted as a free spirit, a genius impassioned and determined to follow his path despite prevailing social and institutionalised norms (Radcliff 63, 67). According to Barker et al. (9), this transformation towards an increased valorisation of imagination, originality, creativity and self-expression needs to be considered within the confines of broader economic and political developments, such as the Industrial Revolution and the French Revolution. It is also argued that the historical privileging of subjectivity is related to capitalism's base in "the autonomy and self-fulfilment of the humanist conception of the individual" (Bennett 16). Pierre Bourdieu (304) stresses the importance of a social development, i.e. the presence of an avant-garde bohemian audience in 19th century Paris that offered the artistic innovators the necessary "symbolic protection", to explain the advent and development of the avant-garde artists.

The literary author too, as a distinct type of artist, has a long tradition of being accredited with an authoritative position in society. More precisely, the notion of the author evolved from the ancient Greek view of the poet as prophet to the medieval belief in the author as intending to enhance humanity through knowledge towards a paradoxical notion – which emerged during Romanticism – of simultaneously understanding the author as an unconscious, depersonalised and absent genius and as a self-expressive and social person creating for a larger audience. According to Andrew Bennett, this paradox is intensified in modern and postmodern societies in which the author is both declared death and desired for at the same time (36-66). Contemporary notions of the Romantic author, however, often refer to an ideal of autonomy (ibid. 71), grown from a seemingly ever-present "social desire for a transcendent originary unity", Bennett (35) concludes.

Mediating artisthood

It can be argued that artists, to a certain extent, add to the maintenance of this authoritative image. This is exemplified in the way the advent of mass media influenced how artists reflected upon their life and work. For instance, it is argued that several artists developed a doctrine of "*l'art pour l'art*", a so-called pure art (Lovejoy 34). This was a way for various artists to distinguish themselves from the popular and mass-produced cultural forms, which they regarded as aesthetically inferior (Bruder 162-169). The artist, thus, tried to maintain a unique position in modern society. This differentiation crystallised in various forms whereof dissociating from popular culture was only one. In light of their self-affirmation and social distinction, artists also reacted against the bourgeoisie. This meant a breakaway from the classic Academy's influence via the "invention" of both the heroic, rebellious artist and the doomed, tortured artist (Bourdieu 166). Furthermore, some artists perceived experiential knowledge as a necessary prerequisite to understand art. Therefore, they argued that this could not be left to the unattached art critic or to the ignorant art public (Bruder 170-172). Finally, artists were also involved in an internal struggle over the conditions that allowed someone to enter the artistic field and to assume the status of true artist. At stake in this battle was the defence of the established order, the monopoly of artistic legitimacy and the authority to call oneself an artist (Bourdieu 270-273).

However, over the years the proliferation of electronic technologies motivated artists to make use of mass media technologies, such as television, to reach a broader audience, although the related loss of control for the artist did not always pass without a struggle (Lovejoy 122). In addition, in the 1960s in the United States young artists no longer acknowledged their working place to be an isolated "sanctuary", yet defined it as a "bustling factory" in which technology occupied a special place (Jones 357-361). Nevertheless, the inclusion of "others" in the artist's studio did not imply an erosion of the artist's creative superiority. The artistic idea still arose from the head of the artist, even though there was more than one individual constructing the work of art (ibid. 371-372).

This paradox also appears in the way literary writers have been puzzled over adapting their standards of what is art to the public's wishes. Refusing to admit to this adaptation does not, however, necessarily imply that the writer wants to be locked up in an ivory tower. As Leo Lowenthal (24-25) mentions, the writer Goethe, for instance, stated that it was the writer's task to raise the intellectual level of the audience and to be concerned with the outside world. Yet, at the same time, Goethe also emphasised the need for complete freedom for the writer, "following only his inner voice" (ibid. 25), to be able to bring this task to a favourable conclusion as well as a refusal to create inferior art that "aims only at entertainment" (ibid. 21). Writers, hence, had to find a compromise between their own standards of art and the divers tastes of this new, modern audience (ibid. 95).

In a more recent conversation the Belgian artist Angelo Vermeulen and the Belgian philosopher Antoon Van den Braembussche mention that at present these Romantic conceptions of the artist still slumber and that artists, even today, are not very eager to open up their creative processes to the world (Van den Braembussche and Vermeulen 86, 233). Hence, it seems that the abovementioned paradox of the artist understood by some as positioned outside society and by some to be engaged in society still holds sway in contemporary society.

Blogging and (de)mystification

In our research, we question what interactive media might mean in relation to this paradox. We understand the weblog as an interactive medium, hereby building on Spiro Kiousis' (378) definition of interactive media based upon the presence of distinct properties of technology, the communication context and users' perceptions of interactivity. It can be argued that weblogs empower literary writers to mediate and to take control of their self-presentation (boyd). At the same time, it can be claimed that the traditional position of the media audience is also potentially altered, for the audience is now given the ability to engage in a more active way in the production of this type of symbolic content (Cammaerts 358). As a result, according to Nigel Whiteley (215-228), today new technologies more than ever have the ability to demystify the image of the artist, for these technologies can be almost all-revealing while being minimally invasive. This idea, however, has been questioned and critiqued. Chris Chesher (web), for one, believes that weblogs actually perpetuate, coexist with and transform conventional authorship. He, therefore, has argued that the so-called death of the author as a result of the interactive and democratic opportunities offered by the Internet is greatly exaggerated: "the Author is alive and well, and has a blog", he says. According to Chesher, the features and conventions of blogs ascribe a certain type of capital to the blog writer that corresponds to the Romantic cultural convention of the author as an individual, mysterious and even magical genius. Hence, Chesher concludes, paradoxically to the predictions of theorists celebrating the liberating potentials of the social web because of its less unidirectional and more interactive qualities, it seems that blogs succeed because they are less innovative than other online forms. He, thus, seems to imply that people still prefer conventional and recognisable communication patterns. A similar view is held by Internet researcher Geert Lovink (20) who states that blog readers' comments are not of equal value to the blog author's postings: "users are guests, not equal partners, let alone antagonists".

When considering blogging writers, we can refer to the work of Piret Viires and Virve Sarapik (353-354), who studied Estonian writers who also create online. These scholars argue that writers want to act solitarily in cyberspace. Rather than being interested in engaging in a dialogue with readers and as a result being influenced when creating, writers mainly intend to get their message across and to stay in control, either by reacting to readers' comments, closing down their blogs or editing already published texts (ibid.).

Jill Walker Rettberg (5-6), focusing on blogs in general, underlines the fact that blogs combine aspects of both dissemination (passive audience) and dialogue (active audience). On the one hand, blogs are published publicly and in a one-way fashion on the Internet. On the other hand, the author and the blog's readers can enter into a two-way dialogue. Hence, blogs are a reflection of solitary, personal ideas as well as of a social conversation. This is not a new development. According to José van Dijck, the traditional, handwritten diary, with which blogs are often compared, is dialogic in nature as well, "obliterating the line between public and private". It is said that diaries and blogs are at the same time a (self-) reflection of life as well as a way of constructing life and connecting with others. The fact that weblogs might be used in either way – as dissemination or dialogue – can be explained from the power or authority they grant the author. According to Jill Walker, only the blog writer is responsible for his or her representation. The writer controls self-presentation and self-reflection via a continuously expanding collection of blog posts "each of which is a micro-narrative or a comment that tends to express an aspect of the writer" (Walker). According to danah boyd, it is this sense of ownership that makes the blog feel like a personal space to the author. This means that a dissociation exists between the blogger, who controls the presence of blog entries as well as of blog comments, on the one hand, and the audience at large, on the other hand. Although the author is willing to engage in conversations, he or she does not perceive the public as all people over all time and space: "This conception of the public is an embodied one … The target audience is not the public at large, but those for whom the topics of discussion matter" (boyd). Blogs are a creation of the self and, thus, reflect selected aspects of the writer's life. This means that blogs veil and unveil at the same time (Walker). The question we can ask, then, is to what extent does the writer veil and, simultaneously, unveil the self through the use of his weblog? The answer to this question should broaden our understanding of the way the concepts of artistic authority and of 'writerhood', as well as related traditional social relations, are constructed within the confines of new, interactive media. For, as danah boyd states, research on blogging must begin "with an understanding that blogs are a medium [through which communication occurs] and this medium has and will continue to shift the communicative and social assumptions that ground everyday life".

Case study Bart Moeyaert: Writer and blogger in residence

Bart Moeyaert is a professional Flemish writer since 1995. He has written several books that were translated into over fifteen languages as well as poems, screenplays and translations. Several of his books have received awards. Since 2003 he has been teaching Creative Writing at the Royal Conservatory of Antwerp (Belgium). In 2006 he was appointed city poet of Antwerp for a period of two years.[1] The blog entries that were written by Bart Moeyaert, and analysed within the context of this research project, cover the period from the first of January 2008 to the 31st of January 2008, which coincided with his last month as city poet of Antwerp. Moeyaert was asked to post a daily blog entry for one month about his life and work on the Villa Kakelbont blog of the Focuspunt Jeugdliteratuur, a Flemish organisation that aims to promote knowledge about children's and youth literature. Since this blog explicitly focused on writers reflecting on themselves and their artistic practices, and since its timespan was limited, we considered it to be a valuable pilot case. Moeyaert published 42 posts in total. These were written in Dutch, each time accompanied by a picture and almost always comprising multiple paragraphs. Blog readers were allowed to submit comments to every post. These comments, however, were not published on the blog directly. One had to click the comments button underneath each post to view the comments.

In this article we solely focus on our analysis of the writer's blog posts. The aforementioned theoretical perspectives inspired the coding framework that guided this qualitative analysis. Yet, the codes were developed during the analysis, allowing an open reading. We build on the methodology of thematic narrative analysis described by Catherine Kohler Riessman – keeping in mind that we study blog narratives rather than interview narratives – as a type of interpretative analysis of subjectivity, identity and representation (Kohler Riessman 2-5). The underlying research questions that directed our analysis were aimed at understanding how the writer constructs reflections about his life and work and about his relationships with his offline and online audience and with society in general, through the use of his weblog. We did not look for factual information, rather for underlying narratives and conceptualisations. Moreover, we interviewed Bart Moeyaert in October 2010 to get an insight into the way he experiences his blogging activities, then and now. This allowed us to find plausible explanations for several observations that came out of the narrative analysis.

1 Information retrieved from http://www.bartmoeyaert.com on 21 January 2010.

Distinction, social embedding and individual agency

Based on the findings of our case study, it appears that the relationship between the writer and his audience, as shaped through the blog posts, is mainly characterised by distinction and mystification. This means that while blogging, Moeyaert reinforces the mythical image of the writer in multiple ways. Several of his blog posts suggest a notion of the great artist and characterise the artist's materials with a semblance of mystique. For example, he describes his desk as a "neo-gothic monument from a convent" (30 January 2008).[2] Furthermore, it is clear that Moeyaert attaches great importance to his power of decision and to the way he regulates his life and working practices. He states that the writer "rows against the current" (31 January 2008) and, hence, does not like to be forced into doing something a certain way. Moeyaert describes himself as having a wide outlook on society and as being open to the world as it is. He therefore expects to be treated alike and be allowed his freedom. Moeyaert stresses that an artist should remain himself and pursue his ideals. This, however, implies a certain distance between the writer and society. Moeyaert mentions that "life is safer between the walls of my workroom" (8 January 2008) and that "between the writer and the reader there is always time, and time is safe. Time is distance" (8 January 2008). Hence, he seems to imply that there is a place – physically and intellectually – that the writer does not want to share with the audience. This type of isolation and quiet is said to be essential for the writer to be able to create: "I have been there for two years [he refers to his time as city poet], approachable, attainable, willing, but now not for a while. I am returning to myself" (31 January 2008). These observations were reinforced during the interview. Here, he mentioned the importance of being outside society, i.e. not being available for some time, when working on a new book. On the contrary, he emphasised his responsibility of being available for the public, yet only when not creating. Since, in essence, Moeyaert says, the writer has to create alone: "you are sitting between four walls and basically you have no friends. You have to do it alone" (Moeyaert, interview 20 October 2010).

Furthermore, in the blog posts, Moeyaert seems to reflect on the relationship between artists and society in terms of two groups of people. There are artists and there are those who are not artists. These two groups do not really know each other: "Neither young nor old will ever fully understand what a writer does. You don't need to explain why chocolate is good. To prove that a book can be good is difficult" (17 January 2008). Similarly, as a writer he experiences performing in front of the public as sometimes confusing: "I can't make head or tail of it. It is a

2 These citations originate from Bart Moeyaert's blog postings and were translated by the author of this article. The blog posts were originally written in Dutch.

mystery to me what makes a lecture become pleasant, or not" (17 January 2008). Moeyaert's blog writings, hence, indicate that artists belong to a distinct social group. As a result, he values artists who refuse to compromise on their freedom and authority and who want to be taken as they are. Moeyaert even describes one artist as "one of my heroes" (16 January 2008) and, thus, stresses the impression of a hierarchical relationship between the artist and the beholder, between the unreachable artist, whom he describes as "Mister Wilfulness Himself" (27 January 2008), and other people. In the interview Moeyaert also underlined that in the first instance he writes for himself, to shape himself, to find himself, to complete himself. Only in the second instance does he write his books for an audience, "otherwise I would not publish it" (Moeyaert, interview 20 October 2010).

Moeyaert also blogs about the least pleasant aspects of being a writer. These aspects are mainly connected with the relationship between the artist and the audience at large, for instance in the case of public lectures that can be very time-consuming: "I sometimes talk more about writing than actually writing something, and sometimes that is really unpleasant" (10 January 2008). Although he acknowledges that he wants to reach people, he is pleased with the help of others to take care of some of the accompanying necessary worries of realising artistic plans. Furthermore, he underlines that a lot of these public activities, such as giving press interviews for television, do not always represent him the way he actually is: "I read in my own volume, which to me is even more awkward than the fact that this volume is kept in the drawer of my cupboard, but I know, if it is well edited the viewer will like it" (22 January 2008). As a result, in several blog posts Moeyaert suggests that there should be a certain distance between writers and their audience, not in the least to avoid "pleasing the audience", which he characterises as "exhausting, for it appears to me to be absolutely tedious to constantly have to make choices that are not mine; it would erode me" (29 January 2008).

In the blog Moeyaert also ascribes a distinct social embedment to the writer, whom he considers to stand beside society. This position allows the writer to observe, he suggests: "I found it meaningful that I would look from behind my desk to all those passing lives behind those windows" (30 January 2008) and to confront society with his observations: "I show how rude rubbish can sound" (24 January 2008). The ability to rouse and to move the beholder, he believes, constitutes the power of art. During the interview he bears out this point of view by stating that "the word has power" (Moeyaert, interview 20 October 2010) and that people are in need of someone who offers them a guideline. This, he believes, is his "responsibility" (Moeyaert, interview 20 October 2010) as a writer, yet at the same time he emphasises that he does not feel entirely comfortable with being put on this pedestal. Though he truly believes in the value of literature as well as in his task as a writer to offer the audience his "major truths" (Moeyaert, interview 20 October 2010), he states that he is not willing to adapt his work to this aim.

A great deal of Moeyaert's blog posts are directed at himself. These seem to suggest that the writer is a unique personality rather than being like everybody else. For example, Moeyaert tries to offer the readers of his blog posts insight into what it means to be a writer. In these instances, however, it seems that the writer is aiming for a strengthening of the mystification of artisthood rather than demystifying it: "I mould and I mould and I mould, until I believe I cannot mould it any better, and if someone shakes their head ignorantly when seeing what I have done, then all I can say is that I couldn't do it any better at that time" (8 January 2010). In the interview we can find a plausible explanation for this paradoxical situation. According to Moeyaert, his life and work are very much interwoven. This interrelationship is public. Contrarily, particular parts of him are private and are not intended to be shared. In this context he refers to elements that solely focus on his personal life, for instance "what happens in the garden, the kitchen, the bedroom" (Moeyaert, interview 20 October 2010), or that solely focus on his work, such as the creation process of a new book. So, while in his blog he mentions very personal characteristics, which to a certain extent unveil the mystery that surrounds the writer, this at the same time happens in a very controlled and thought-through manner. He decides what to present about himself and how to present it, always focusing on what he calls "the oeuvre" (Moeyaert, interview 20 October 2010) of both his life and work. He wants to show the different parts that, when put together, make Moeyaert, the writing person. Not Moeyaert when writing, nor Moeyaert the person. These separate aspects remain part of the mystery. Moeyaert acknowledges that controlling his blog in this way offers him "comfort" and "peace" (interview 20 October 2010).

Interactive relations

Notwithstanding the emphasis Moeyaert puts on the writer's distinctiveness, some blog posts point towards the presence of a different narrative. Although this narrative is present only to a minor degree, it demystifies the image of the solitary or distinguished writer by elaborating on the writer's relatedness with others instead of his or her distinctiveness.. This narrative, thus, seems to refer to the artist having a more open, direct and personal attitude vis-à-vis the public, in which he or she is open for a more equal and two-way dialogue. This refers to what Immanuel Kant called a *sensus communis*, a common sense of life (Van den Braembussche 152-158), a connection between writer and reader. For instance, some blog posts are used to reflect on the pleasure the writer gets out of the encounters with readers who are, just like him, open-minded (17 January 2010). In addition, Moeyaert seems to get much satisfaction from the direct encounters with a public that responds by showing him that what he is saying affects them: "The girls and boys don't know how amazing I find them. Their faces brighten. They are all eyes and ears. When suddenly one of the boys takes his chair and comes sitting very close to me I am a bit off my stroke, because it appears that what I am telling finds mark" (10 January 2008). In other words, Moeyaert's blog posts show that he values interactions that reflect a type of connectivity, of understanding between the writer and like-minded people. In the interview he emphasises that these types of encounters stimulate him to "learn" (Moeyaert, interview 20 October 2010) and to reflect. He elaborates on how the blog medium and other online interactive applications have enhanced his perception of this relatedness between the writer and a distinct group of people. For, on the one hand, some blog readers' reactions have made him more aware of the fact that he is not alone, that other people too create and reflect and want to share their contemplations of life. On the other hand, he enjoys the fact that he can make these people feel connected with him in a like-minded way. He, hereby, distinguishes rather thoroughly between comments from the audience at large that he experiences as mostly shallow utterances, and comments from this distinct group of like-minded spirits that are of "interest" (Moeyaert, interview 20 October 2010) to him.

Constructing writerhood online

In this article we focused on one specific case. Yet, by combining both a thematic narrative analysis of the blog posts and an interview with the writer, we believe that this case study allows us to better understand what blogs can mean for both the writer's self-presentation and his or her relationship with the audience. We can conclude that the way Bart Moeyaert makes use of the weblog medium does not differ greatly from the way writers have reflected about and presented themselves offline. We noted that the blog posts show a unification of both mystification and demystification of artisthood. This means that the paradox, which was elaborated on through the historical-theoretical framework, does not dissolve. Both main narratives, being the notion of the mysterious writer-genius and the notion of the writer creating for and in interaction with an audience, appear in the blog. However, these narratives are not present to an equal extent in the blog posts, and what is more, the analysis showed that the blogging author constructed a complex image, also focusing on his personal self and on personal encounters with readers. In the context of this case study, traditional narratives emphasizing distinction and the author's agency dominate the blog posts, whereas what we could call an interactive narrative of writerhood appears the least. Taking the dominance of these 'old' narratives into account, it can be concluded that the writer's weblog posts emphasise the literary author's differentiation from others rather than his or her connection with others. This means that notwithstanding the newness of the weblog as medium, which is related to the possibility of publicly disseminating personal meanings and symbolic constructions as well as to the opportunity of responding to and dialoguing with readers (Walker Rettberg 5-6), old narratives and traditional ways of thinking seem to survive on the weblog. This also shows through in the interview wherein Moeyaert mentions several times that the blog medium offers him "comfort", since it allows him to better maintain a "healthy balance" (Moeyaert, interview 20 October 2010) between giving time to the audience and taking time for himself. He is convinced that by posting on a regular basis, his responsibility to constantly be there diminishes. However, at the same time, we did observe a distinct change in the writer's perception of his relationship with the public. The interview revealed that the fact that personal encounters with the audience were mentioned in some of the blog posts, relates to an alteration of the notion of writerhood on a distinct level. Namely, as Moeyaert states, online reactions that have depth make him aware of the fact that he can feel connected with a specific group of reflective people from whom he can learn and that this "indirectly" (interview 20 October 2010) changes him as a writer. What is new, then, is that the writer himself to a larger extent perceives, experiences and expresses the relationship with his offline and online readers through his blog posts. Summing up, the weblog as a 'new' medium is not used here to simply demystify the literary author's traditional authority and image, even though personal encounters with the audience are integrated in the author's online self-conceptualisation.

As a next step in our analysis, we believe it to be crucial to try to explain the writer's blogged self-conceptualisation of both distinction and connection. Several aspects can be taken into account. For instance, the context of the weblog might be important. The blog posts that were analysed in this case study were published on an organisational weblog, which had a specific aim for Moeyaert, namely writing down his experience of his last month as city poet of Antwerp, instead of on the author's personal weblog. In addition, the cultural background and literary tradition of which the writer is a part might explain why in this case the mythical and distinct, individualised image of the writer dominated the blog posts. For example, in the interview, Moeyaert mentions that, contrary to the Netherlands, in Flanders writers are generally not the ones standing on the barricades to fulfil a social role. Finally, an analysis of readers' blog comments can help to understand how writerhood becomes conceptualised through these online, interactive encounters.

References

Barker, Emma, Nick Webb, and Kim Woods. "Historical Introduction: The Idea of the Artist." *The Changing Status of the Artist.* Eds. Emma Barker, Nick Webb, and Kim Woods. New Haven; London: Yale University Press in association with the Open University, 1999. 7-25. Print.

Bennett, Andrew. *The Author.* London; New York: Routledge, 2005. Print.

Bourdieu, Pierre. *De regels van de kunst. Wording en structuur van het literaire veld.* Trans. Rokus Hofstede. Amsterdam: Van Gennep, 1994. Print. [Bourdieu, Pierre. *The Rules of Art: Genesis and Structure of the Literary Field.* Cambridge: Polity Press, 1996.]

Boyd, Danah. "A Blogger's Blog: Exploring the Definition of a Medium." *Reconstruction: Studies in Contemporary Culture* 6.4 (2006): n. pag. Web. 1 Mar. 2010. http://reconstruction.eserver.org/064/boyd.shtml.

Bruder, Anne. "Constructing Artist and Critic. Between J. M. Whistler and Oscar Wilde: "In the Best Days of Art there Were no Art-Critics"." *English Literature in Transition, 1880-1920* 47.2 (2004): 161-180. Print.

Cammaerts, Bart. "Critiques on the Participatory Potentials of Web 2.0." *Communication, Culture & Critique* 1.4 (2008): 358-377. Print.

Chesher, Chris. "Blogs and the Crisis of Authorship." *Blogtalk Downunder 2005*. Sydney, 2005. Web. 9 Feb. 2009. http://incsub.org/blogtalk/?page_id=40.

Jones, Caroline A. *Machine in the Studio. Constructing the Postwar American Artist.* Chicago; London: The University of Chicago Press, 1996. Print.

Kiousis, Spiro. "Interactivity: A Concept Explication." *New Media & Society*, 4.3 (2002): 355-383. Print.

Lovejoy, Margot. *Digital Currents: Art in the Electronic Age.* New York; London: Routledge, 2004. Print.

Lovink, Geert. *Zero Comments: Blogging and Critical Internet Culture.* New York: Routledge, 2008. Print.

Lowenthal, Leo. *Literature, Popular Culture, and Society*. Englewood Cliffs, NJ: Prentice-Hall, 1961. Print.

Moeyaert, Bart. Personal interview. 20 Oct. 2010.

Radcliff, Matthew. "Absentminded Professor or Romantic Artist?: The Depiction of Creativity in Documentary Biographies of Albert Einstein." *Journal of Popular Film and Television* 36.2 (2008): 62-71. Print.

Rettberg, Jill Walker. "Blogs, Literacies and the Collapse of Private and Public." *Leonardo Electronic Almanac* 16.2-3 (2008): 1-10. Web. 1 Mar. 2010.

Riessman, Catherine Kohler. *Narrative Analysis.* Newbury Park, CA; London; New Delhi: Sage Publications, 1993. Print.

Van den Braembussche, Antoon A. *Denken over kunst. Een inleiding in de kunstfilosofie.* 4th ed. Bussum: Coutinho, 2007. Print. [Van den Braembussche, Antoon A. *Thinking Art: An Introduction to Philosophy of Art.* [Dordrecht]: Springer, 2009.]

Van den Braembussche, Antoon A., and Angelo Vermeulen. *Baudelaire in Cyberspace: Dialogen over kunst, wetenschap en digitale cultuur.* [Dialogues about Art, Science and Digital Culture.] Brussel: ASP (Academic and Scientific Publishers), 2008. Print.

van Dijck, José. "Composing the Self: Of Diaries and Lifelogs." *Fibreculture Journal* 3 (2004): n. pag. Web. 1 Mar. 2010. http://three.fibreculturejournal.org/fcj-012-composing-the-self-of-diaries-and-lifelogs/.

Viires, Piret, and Virve Sarapik. "Solitude in Cyberspace." *Transforming Culture in the Digital Age. International Conference in Tartu 14-16 April 2010.* Ed. Agnes Aljas et al. Tartu: Estonian National Museum, Estonian Literary Museum, University of Tartu, 2010. 351-355. Web. http://dspace.utlib.ee/dspace/bitstream/handle/10062/14768/transform_book.pdf.

Walker, Jill. "Mirrors and Shadows: The Digital Aestheticisation of Oneself." *Digital Arts and Culture Copenhagen 2005.* Copenhagen: IT University, 2005. Web. 1 Mar. 2010.

Whiteley, Nigel. "Seeing *What*, *How* and *Why*: The *ARTnews* Series, 1953-58." *Journal of Visual Arts Practice* 6.3 (2007): 215-228. Print.

Joke Beyl is a PhD Student and Teaching Assistant at the Department of Communication Studies, Free University Brussels. She is affiliated with the research centre iMinds-SMIT (Studies on Media, Information & Telecommunication). Her PhD research focuses on the construction of writer identity in the encounter between blogging writers and commenting readers.

Life-Publishing on the Internet – a Playful Field of Life-Narrating

Sari Östman

PhD Student
Digital Culture, University of Turku
Department of Cultural Production and Landscape Studies
PL 124
28101, Pori
Finland
sari.ostman@utu.fi

1. Introduction

Life-publishing is a new concept that I have been developing since 2007. It includes different ways of relating one's everyday life on the Internet. I am currently studying life-publishing as a field of autobiographical practice. My material triangulation includes 30 thematical narratives, 18 interviews and over 70 internet publications that are produced by 30 Finnish informants who have signed up for my research on a web survey. In this paper, the material consists of 15 informants who had written their narratives by the autumn of 2010. Ten of them had sent me their texts during the pilot period in 2008-2009. The remaining five narratives were more recent, produced in the winter of 2009-2010.

Ten of the 15 informants have given me their permission to follow their publications. Since some of them publish their lives on many different forums, the total of the web contents for this paper is 22 publications. They include blogs and micro-blogs, communal sites (e.g. Facebook) and homepages.

Three of the informants are men, the rest are women. This does not necessarily mean that life-publishing is 'feminine' by nature, although it seems possible that women are more likely to recognise themselves as life-publishers. The ages vary from 21 to 61 and women between the ages of 21 and 34 are a clear majority.

In my previous studies I have articulated three basic dimensions or elements: life-publishing is *reflective*, *performative* and *playful*. These aspects are strongly related and form the foundation of life publishing as an everyday activity. Reflectivity means 'I', the intimate personal life; performance relates to 'you', the audience; playfulness represents 'us', the sociality of life publishing. In this paper, I concentrate on one important element that includes the other two. I will ask: How is life-publishing playful according to a) thematical narratives and b) publications of 15 informants? First, I will define the concept of life-publishing. Then, I move on to analysing the thematical narratives and publications. Finally, I will briefly sum up my findings. When quoting my informants, I will mark thematical narratives with a mention of "TN" and the publications, which consist mainly of blogs and Facebook profiles, by "B" and the publishing date for a blog, "FB" and the publishing date for Facebook. Other publications are not directly quoted. Before that, it is important to define what I mean by 'playful' in detail. Johan Huizinga – probably the best known researcher of play – has articulated certain points that define 'play': play is free but it still has rules, sometimes even very strict ones. Play is seemingly meaningless even with its rules and aims; it is also bound by time and space. It is organised and has excitement. (Huizinga 16-21.) On the other hand, Riitta Hänninen, a Finnish cultural anthropologist, has studied play as a concept and phenomenon. She emphasises Hans-Georg Gadamer's definition of play (spiel): the *experience* of a certain activity as play is what makes it play. Hänninen points out that anything can be turned into play by sending the meta-message "this is play!". (Hänninen 49-50) Roger Caillois's idea about the notion of "as if" as a meta-rule for play, supports Hänninen's thinking (8).

These are the basic guidelines for what I understand by 'playfulness' in general. However, in studying Internet life publishing, I will also strongly build my understanding of playfulness on the idea of Georg Simmel's *sociability*. Simmel has written (112-131) about affability. In his opinion, sociability in its widest sense becomes apparent in "everything that we call play" (114). That is, in everything that can sociologically be seen as play (122). Simmel notes that in sociable communities, being tactful is extremely important – sociability is a form of play in which we act *as if* (compare Caillois 8; Hänninen 49-50: both see "as if" as one of the most important indicators of play) everyone is equal and especially respectful towards one another (118, 122). The tactfulness affects the form and style of actions. This idea also appears to be strong among life-publishers. They place much value on the quality of the content they produce and rely heavily on netiquettes and media education speech as well as other control mechanisms that guide their behaviour online.

As Simmel writes, sociability is contradictory as it expects certain personal touches while, at the same time, not allowing for the most personal and intimate aspects of life, character and mind (118). I have come to quite similar conclusions when studying the most common discourses concerning life-publishing. Simmel argues that people often complain that sociability is superficial (129); this view fits well with the most common life-publishing discourse (Östman, "What is a Good Life Publisher Made of – and Why?"). This discourse about social media activity being narcissistic self-branding is also criticised by Multisilta, Mäenpää and Suominen. Applying Simmel's notion to contemporary media culture, Multisilta et al. have coined the concept of *media sociability* (10). When studying this further, I found that Simmel's sociability applies to the playfulness of Internet life-publishing. Similarities between the notions of sociability and playfulness could be discussed further, but in this article they are used interchangeably.

Simmel also notes that performance and narrativity are among the key features of sociability; this is also the case with life-publishing. Jill Walker Rettberg has studied the narrativity of blogging, dedicating a whole chapter to narrativity and identity play in her *Blogging* (111-126). She writes about the flirtatious and intimate tone of voice that diary bloggers may use. In Rettberg's words, they "play a ... game of peek-a-boo, showing but not showing all". (120) She notes that most blogs are at least partly self-representational, and thus, a form of life writing. The character of a typical blog narrative she describes coincided with my findings: various fragments create a larger narrative when read together. Rettberg calls this ongoing and serial narrative. When considering identity play, she reminds us that most blogs are based on real individual's thoughts. I assume that by this she means that even though 'fake' life publications do exist, they are marginal. Moreover, honest, diary-like blogs, gallery sites and such include identity play, which becomes visible most clearly in what I will later in this article call "self-oriented play". Identity play is also present in performance play and social play (for more see below).

Finally, I must note that when we speak about the playful experience, we are actually speaking about more than life-publishers' experiences of being playful. Since the publishers do not often explicate their playfulness in so many words, the experience is mainly constructed in analysis. Therefore, in this study, the experience of playfulness – or media sociability – is the researcher's interpreted experience of the informants' original experience.

2. Life-publishing as a concept

Life-publishing is a hybrid form of lifewriting. I developed the notion while studying life blogs by Finnish women. Life-publishers represent aspects of their lives publicly through media, in this case the Internet. Life-publishing draws from other autobiographical traditions but brings new aspects to the field. The levels of publicity and social intimacy are freely chosen by the writer. By social intimacy I mean the tendency to describe everyday life in a seemingly open-hearted and intimate manner. Communal sites, photo galleries, video streams and discussion forums are all forms of life publishing. However, written blogs and their predecessors, personal home pages, are still probably the best known forms of life publishing. While studying home pages, media researcher Susanna Paasonen has found certain similar elements to my findings, such as playfulness, family intimacy, everyday life and celebration. According to Paasonen, personal homepages are different from professional ones because the former are a kind of identity work, "peek[ing] into private spaces and events". ("Figures of Fantasy: Women, Cyberdiscourse and the Popular Internet" 94-126; "Tervetuloa henkilökohtaiselle kotisivulleni!" passim.) Today, blogs attract wide interest among scholars. For example, media researchers such as Geert Lovink, José van Dijck as well as informatics researcher Susan Herring have studied blogs. Lovink is especially critical of blogging and other "John and Jane Doe's" practices of web publishing; he has even referred to them as an Internet apocalypse that eventually leads to complete decadence of the Internet contents and cultures. Lovink's view in the article in question is meant to be ironic, but it indicates a very common discourse that surrounds life-publishing. (Passim.)

In light of my research material, this discourse is hard to relate to. Publishers reflect their activity in their texts, explaining to themselves as well as their audiences why and how they tend to describe their intimate everyday lives on the web. They justify their actions, control and re-control their contents consciously:

> I have cut down talking about especially personal things in my blog, because while writing under my real name I want to keep some things my own. Now, however, I feel that writing will help and sooth, although sharing my personal sorrow publicly does have its downside. But I just feel so bad, so I share it here to get my sorrow structured at least a bit. (Woman, 27 years old, B Sept 25, 2009. Translated by SÖ.)

Why is it so important in the first place to recognise life-publishing as its own field of practices, and how should it be situated in life narrating traditions? Since written blogs are the best known of these practices, their diary-like nature has been questioned very little. This may be caused by the fact that women have for years been the majority in written life publishing, at least in Finland. Diary writing in the *journal intime* form (on this term and dividing journal into *intime* and *externe*, see Gusdorf 39-42) has a traditional nature as a feminine practice of self-expression (e.g. Leskelä-Kärki 77; Makkonen 236). However, male life-publishers exist, and the percentages of men and women seem to be equalising. Jill Rettberg notes that most blogs, even goal-oriented ones, are self-representational at least to some extent. Therefore, it is important that life-publishing is not seen only as the web diaries of (young) women. Life-publishing has a much wider extent, both in contents and technologies. It includes photo galleries, communal sites, professional performances and much more. It tells about one's actions, observations and feelings (Östman, "What is a Good Life Publisher Made of – and Why?").

The common understanding of 'diary' in the 20th and 21st centuries is that diary is something highly intimate that is usually kept in secret. In this, life-publications differ from traditional diaries: life-publishing – as the term suggests – is fundamentally public. The publisher has freely chosen the Internet as the open forum for performing her or his life. The level of publicity can be chosen and re-chosen as many times as the publisher wants to; in traditional publications the choice has always been someone else's to make. In addition, the level of publicity is constantly under re-definition and discussion. Life-publishing also has similarities to other autobiographical practices such as actual autobiography or memorising speech (autobiographical practices have been studied in great amount: see, for example, Fingerroos and Haanpää; Kosonen; Saresma or Ukkonen). However, it cannot be brought back to any of them in its entirety. Life-publishing is almost or completely concurrent and may combine past, presence and future as a seamless entity in one single post. It is fragmentary, but the fragments can be collected like puzzle pieces: they constitute a whole (see also Rettberg 111, 115).

These new practices are aided, yet not caused, by the technological change in which computers, the Internet and mobile devices have become more and more common. The change is also cultural; people are starting to consider Internet life-publishing and content sharing natural activities that do not constantly need to be questioned and defended. (On cultural change in Internet practices in the Finnish context, see Saarikoski et al.)

Life-publishers themselves have compared their activity to diaries as well as letters or real-time autobiographies. In earlier theme writings, and especially in written blogs, I have also come across the comparison to memory speech. (See "Avaudutaan arjesta. Suomalaisnaisten Internet-elämäjulkaiseminen ja maskuliinisuuden esittämisen keinot" 25-40; "Elämäjulkaiseminen – omaelämäkerrallisten traditioiden kuopus".) Comparisons and similarities show that life-publishing draws strongly from the tradition of autobiographical practices. This autobiographical nature, this everyday-ness, also creates a dilemma. Anne Kaun quotes Roger Silverstone's words about everyday life: she reminds that it does not fit the strict definition of play. Kaun therefore understands play as something that is secluded from the everyday, from routine. (95)

Life-publishing cannot, however, be studied in a similar way to traditional everyday life writing. Every publication is intertextual in a way that, for example, a diary, letters and an invited autobiography from one person might be when examined together. The consciously chosen publicity proves that the activity is meant for audiences at least as much as for the publisher. These qualities give reason to suggest that life-publishing is a new subfield among autobiographical practices: a playful performance, yet intimately reflective. Life-publishing is a sociable and therefore playful media activity.

3. Playful reflectivity: Thematical narratives of the publishers

Fifteen informants had sent me their thematical narratives by February 2010. The term *thematical narrative* is based on the concept of thematical writing. Thematical writing means that the researcher gives informants a theme or themes on which they write. The form is very free and the questions, or rather just topics, may be followed quite loosely. My method, however, widens the concept: a thematical narrative can be a traditional written text, but also a video, comic or anything the informant feels to be a natural way of expressing her or himself. The only limit is that the researcher must be able to analyse the 'text'. The fifteen narratives in this paper, however, are traditional texts – all but one. One informant sent me a PowerPoint presentation with lots of photographs and almost poetical text fragments.

The first part of my question is: How is life-publishing playful according to thematical narratives? I have analysed fifteen narratives and the results are somewhat surprising. I had a hypothesis that play in several forms would have quite an important role in life-publishers' narratives. Instead, the word 'play' only appeared in one narrative, with almost a negative tone. A young woman describes her views on Facebook, saying:

> It is mostly a playground for grown-ups, in which you can send foolish messages and Tetris. It is not of any benefit, actually. (Woman, 28 years old. TN. Translated by SÖ.)

Three of the informants did not give permission to observe their publications and two of them did not have any. The latter two had many opinions about life-publishing in general, which is why I have included them as informants. Two of the first three did not recognise themselves as life publishers, although they had, for example, a blog or a Facebook profile. These informants were highly critical of the whole life-publishing activity: it was referred to as being revealing, open, un-thinking – 'low culture'. These informants considered, for example, work-related blogging, having a profile on LinkedIn or strictly limited journal blogging as what I in this context call 'high culture': smart, rational, pondering, controlled – in a way more acceptable than life-publishing. This standpoint was to some degree also taken by those who recognised themselves as life-publishers:

> I give a 'smart' picture of myself [in my publications]: I try to be a thinking person who has good, intellectual hobbies. I do not talk about my alcohol usage, eating, housekeeping, work stuff or my current unemployment. (Woman, 56 years old. TN. Translated by SÖ.)

Playfulness is implicitly present in three narratives (a young man and two young women, 21-35 years old). These informants do not mention play or playfulness directly but write about life publishing in a light manner. 'Light' here should not be mistaken for superficial. It should rather be seen as 'serious lightness' – the manner of speech is airy, yet thoughtful.

> As a life publisher I raise myself up into the spotlight, I beg (and get) attention, I get to fulfil my passion for writing and it gets encouraged, too[.] I enjoy writing there… I just truly love my blog. (Woman, 27 years old. TN. Translated by SÖ.)

In this quote, the informant portrays herself as a web publisher. She recognises herself as a performance creator and describes her emotional relationship with her blog. Even though thoughtful and reflective, the tone of this 27-year-old informant is lighter than the earlier quote. The 56-year-old woman seems to write about giving herself a 'smart' self image without irony: she emphasises that she does not tell her readers about her personal life as such, but instead expresses her views on life in a more intellectual way:

> I talk about nature-related events, for example things that happen on the bird feeding place, the frog wedding in the spring, the animal prints that have appeared in the yard during the night, often with photos… My own actions I describe only using suggestion. (Woman, 56 years old. TN. Translated by SÖ.)

One of the three male informants has implicit playfulness in his narrative. He writes about "documenting amusing stories" and "writing down the most humorous inside jokes I made up with the guys". His overall tone is also lighter than the tone of the other two men – or the majority of the women for that matter. All in all, sex is not a major factor in the informants' attitudes towards life publishing.

The man who documents amusing stories and jokes does not blog much about his personal life, contrary to the other two male informants: they describe their life-publishing as journal-keeping and self-therapy. One of the male informants says that his blog writings are his "most genuine, most owned thoughts". The two other male informants seem – according to their theme writing – to take their blogging quite seriously. By this I mean that the tone of their writing is anything but playful: one emphasises the blogger's responsibility for 'blogging well' while the other puts much weight on his privacy. This kind of discourse suggests that these bloggers are having "serious fun", as Jenkins describes the combination of politics and popular culture (e.g. 208-209). This certain seriousness is also visible in female informants' writings. It has to be remembered that this may be caused by the nature of the assignment: in the thematical narratives, the informants have reflected their own actions with more depth than they might do in an everyday situation like a coffee table discussion. Even if a casual reader might see the publications as superficial and unnecessarily open outpourings without any real meaning, the seriousness suggests that life-publishing is significant to the actors and that they give much thought to it.

4. Playful performance: Life-publications

The second part of the question is: How is life-publishing playful according to the publications? I have studied the Internet sites of 15 informants and searched for elements that would suggest a playful approach to intimate web publishing.

'Having fun' (see the definition of play and playfulness in section 1) in life-publications does not necessarily mean that the contents would always be joyful or – as often assumed – superficial. It means rather that the writer ('writer' in the meaning of hybrid media texts) enjoys spending her or his time in life-publishing. The playfulness might not be explicated in publications at all, but it can be read 'between the lines'. However, it becomes visible in layouts, an ironic manner or, for example, when experimenting with different styles of expression. What does this mean among the 15 informants, then?

At first, it must be emphasised that of the 22 publications, 19 are blogs. Their main themes vary to a considerable degree, but certain elements seem to be shared. The blogs are reflective and have an intimate touch with sensitivity, sometimes even sentimentality. Emotions and private thoughts are discussed – not often in detail or with great passion, but always with obvious significance.

> Yes, and at the same time I might like to spend a couple of hours on a café terrace, sipping a latte. Although, really really really with pleasure at least having only [the baby girl] with me and she would sleep this time OR not even [the baby girl] and I could just come and go, hands in my pockets. Now wait… there's a good spot for that in my calendar – right on July 6th… (Woman, 31 years old. B, June 13, 2009. Translated by SÖ.)

Playfulness in these blogs may mainly mean one of two different things: playfulness in the manner of speech and the overall tone of the blog, or playful content, such as this crochet blogger's, who makes amigurumis (Japanese crochet style, see Figure 1):

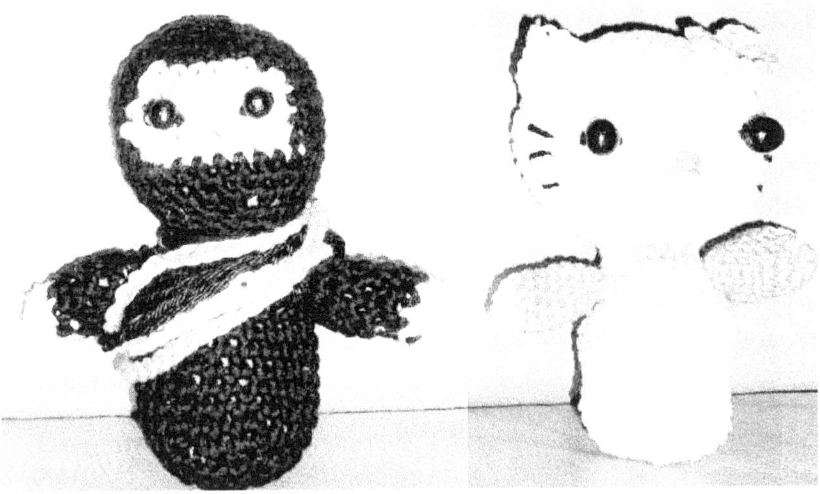

Figure 1: Amigurumis are cute toy-like crocheted figures. Making them is not necessarily play, but the figures themselves, as well as the middle-aged woman's blog about amigurumi crocheting, are playful. Source: http://www.elluntalli.vuodatus.net 20.12. and 23.12.2009

This blogger shares her hobby through her publication. She rarely says anything other than crochet-related things about her life in this blog, but the hobby is a settled part of her everyday life and, therefore, her online activity should be seen as life-publishing. This woman also has two other blogs, both of which are semi-professional. Playfulness in them is more at the level of tone, which even at that level is quite rare. All in all, her playfulness is mainly *self-oriented play*, although she also performs to her audience: this is me as a crocheting artist, what do you think of me?

Other 'content-playful' blogs also have an artistically creative focus: photographs, poems, stories, etc. In Finland, bloggers circulate certain thematic 'games', such as Poem Thursday and Story Monday. The idea is that, in the administrative blog, a theme is given and the participants take a photograph, write a story or a poem and post it. One of my informants takes part in most of these and has a blog through which she posts all her artistic material. She also has a diary blog, in which the tone is quite similar to that of her art: philosophical and thoughtful, sensitive and delicate. Therefore, her manner of speech is not visibly playful. The playful nature of her life-publishing emerges through her participation in creative activities; it could be described as *performance play* – this is me in these photos, this is me in these poems. However, since the point of 'thematical weekdays' is that many people do the same thing from their own points of view, I would also call it highly sociable play.

The rest of the blogs also do not have explicated playfulness, or have it only occasionally. In these blogs, the play is similar to the crocheter's blog: it is self-oriented and intimate and becomes visible in the overall tone – between the lines more than in written words. I will not quote here, since the playful tone is often lost in translation. What remains are mere words that remain dry and uninteresting.

I have given examples of self-oriented, intimate playfulness and performing playfulness. In addition, I mentioned *sociable play*. This can be seen on blogs and homepages, where the intimate tone is emphasised. Here, sociable play happens in memes and challenges. They are circulating question lists or tasks such as "photograph and post the contents of your refrigerator".

> Think pink challenge [in another blog] was so lovely that I felt the need to take part. So, a post in which I model clothes shall finally end up on this blog. Draw a line on the wall, this has finally been seen! (Woman, 27 years old, B, Jan 11, 2009. Translated by SÖ.)

These kind of Internet memes can be seen as play, and they often encourage the publisher to 'loosen up' a bit. The above-quoted blogger often seems to reflect on her life in a deep and thoughtful manner. In participating in the Think Pink challenge, she represents herself in pink slippers and pyjama trousers; she writes more lightly, letting her imaginative hair down.

Although the great majority of the studied contents were blogs, the material also had one Facebook profile, one Twitter micro-blog and a profile that was created for a Finnish social media-related professional site. These three are all from the same informant, a 39-year-old woman, and they explicate sociable play more clearly than the blog material.

This woman has two Facebook sites. She has a personal profile to which I do not have access, so therefore I have studied her fan page, on which she has 255 followers. The person is a known social media 'face', who acts on many related fronts. A fan page in this case does not mean that followers idolise the owner; it is just the easiest way of collecting all her Facebook contacts. In her profile picture this woman wears a pink Stetson, and her overall image is quite feminine. However, the difference between the Facebook fan site and other profiles is that on Facebook, the informant is dressed formally despite the pink Stetson, and her pose is discreet. On the other sites, she plays with a red lollipop, wearing a corset and lots of plastic jewellery. This creates an interesting contrast to the informant's public image as a social media and education professional.

Playfulness in these publications is more easily visible than in thematical narratives. Still it is often delicately humorous and reflective more than lively and comical. The experience of play does not rise from jokes and wild fun; it is implicated in the publication's exterior and overall tone and in the publisher's manner of expression. Therefore, it can be easily seen as tactful media sociability.

5. Summary: Playfulness in life-publishing

In this paper, I have studied Internet life-publishing among 15 Finnish informants. I have asked: How is life-publishing playful according to a) thematical narratives and b) publications? What kind of conclusions can be drawn on the basis of the former sections?

'Play' and 'playfulness' in this study are understood as *experience of play* and as *media sociability*. Since the original experience is not always explicated in media texts, the playful experience is often an interpreted one. In the thematical narratives of my informants, playfulness is only rarely brought up as a meaningful element in life-publishing. However, the details of narratives suggest that the definition of play actualises: life-publishing is voluntary and gives the actor pleasure. Life-publishing also has rules (netiquettes), which is one detail in Huizinga's definition of play, even though it is not as bound to time and space as Huizinga suggested (16-21). In its everydayness life-publishing also shifts away from the strict definition of play. Media sociability, however, suits in life-publishing very well, since Sociability cannot be secluded from our daily routines. And – as Simmel notes – sociability includes everything we can in sociological sense call play.

In publications, playfulness is more visible than in thematical writings. It is rarely explicated in so many words, but is rather to be seen in the layout and other visual choices, along with the overall manner of speech and expressive style. Taking part in memes and challenges is a big part of playful life-publishing. Furthermore, experimenting with styles and methods (e.g. from prose to poetry, from text to video) suggests that life-publishing is seen as a creative, playful activity. For example, Finnish drama pedagogue Hannu Heikkinen has built on Huizinga's ideas when studying 'serious playfulness'. According to him, playfulness might be serious when it is seen as a possibility to study things in a power-free area. This does not at first sound like life-publishing, but it should be remembered that the Internet has traditionally been seen as an unforced, power-free virtual environment. Following Huizinga, Heikkinen argues that play as such is comical neither for the player nor the viewer. (Passim.) This is also how life-publishing should be seen – as serious play with three media sociable aspects: it is self-oriented (intimate), performative and sociable.

References

Sources

Theme narratives from 15 informants between 2008 and 2010.

Internet life-publications from 15 informants (observed in January and February 2010).

Literature

Caillois, Roger. *Man, Play and Games*. Trans. Meyer Barash. Urbana; Chicago: University of Illinois Press, 2001. Print.

Van Dijck, José. "Composing the Self: Of Diaries and Lifelogs." *Fibreculture Journal* 3 (2004). Web. 27 Feb. 2010. http://journal.fibreculture.org/issue3/issue3_vandijck.html.

Fingerroos, Outi, and Riina Haanpää. "Muistitietotutkimuksen ydinkysymyksiä." *Muistitieto-tutkimus. Metodologisia kysymyksiä*. Eds. Outi Fingerroos et al. Helsinki: Suomalaisen Kirjallisuuden Seura, 2006. 25-48. Print.

Gusdorf, Georges. *La découverte de soi*. Paris: Presses universitaires de France, 1948. Print.

Heikkinen, Hannu. *Vakava leikillisyys. Draamakasvatuksen perusteita opettajille*. Helsinki: KVS Foundation, 2004. Print.

Herring, Susan et al. Weblogs as Bridging Genre. – *Information, Technology & People*. 18(2). 2005.

Huizinga, Johan. *Leikkivä ihminen. Yritys kulttuurin leikkiaineksen määrittelemiseksi*. Juva: WSOY, 1984. Print.

Hänninen, Riitta. *Leikki: ilmiö ja käsite*. *Publications of the Research Centre for Contemporary Culture* 76. [Jyväskylä]: University of Jyväskylä, 2003. Print.

Jenkins, Henry. *Convergence Culture: Where Old and New Media Collide*. New York; London: New York University Press, 2006. Print.

Kaun, Anne. "Playful Public Connectivity." *Transforming Culture in the Digital Age. International Conference in Tartu 14-16 April 2010*. Eds. Agnes Aljas et al. Tartu: Estonian National Museum, Estonian Literary Museum, University of Tartu, 2010. 89-99. Web. 15 Nov. 2010. http://hdl.handle.net/10062/14768.

Kosonen, Päivi. *Elämät sanoissa. Eletty ja kerrottu epäjatkuvuus Sarrauten, Durasin, Robbe-Grillet'n ja Perecin omaelämäkerrallisissa teksteissä*. Helsinki: Tutkijaliitto, 2000. Print.

Leskelä-Kärki, Maarit. *Kirjoittaen maailmassa. Krohnin sisaret ja kirjallinen elämä*. Helsinki: Suomalaisen Kirjallisuuden Seura, 2006. Print.

Lovink, Geert. "Blogging, the Nihilist Impulse." *Eurozine* (2007). Web. 27 Feb. 2010. http://www.eurozine.com/articles/2007-01-02-lovink-en.html.

Makkonen, Anna. *Lukija, lähdetkö mukaani? Tutkielmia ja esseitä*. Helsinki: Suomalaisen Kirjallisuuden Seura, 1997. Print.

Multisilta, Jari, Marjo Mäenpää, and Jaakko Suominen. "Mitä on mobiili sosiaalinen media?" *Yhdessä ja liikkeellä. Mobiili sosiaalinen media. Kulttuurituotannon ja maisemantutkimuksen koulutusohjelman julkaisu* 21. Eds. Jari Multisilta et al. Pori: CAT, 2010. 7-19. Print.

Paasonen, Susanna. *Figures of Fantasy: Women, Cyberdiscourse and the Popular Internet.* Turku: University of Turku, 2002. Print.

—. "Tervetuloa henkilökohtaiselle kotisivulleni!" *mediumi.* 2003. Web. 27 Feb. 2010. http://www.m-cult.net/mediumi.

Rettberg, Jill Walker. *Blogging.* Cambridge: Polity Press, 2008. Print.

Saarikoski, Petri, Jaakko Suominen, Riikka Turtiainen, and Sari Östman. *Funetista Facebookiin. Internetin kulttuurihistoria.* Helsinki: Gaudeamus, 2009. Print.

Saresma, Tuija. *Omaelämäkerran rajapinnoilla. Kuolema ja kirjoitus. Nykykulttuurin tutkimuskeskuksen julkaisuja* 92. [Jyväskylä]: Jyväskylän yliopisto, 2007. Print.

Simmel, Georg. *Pieni sosiologia.* Transl. Kauko Pietilä. Tutkijaliiton julkaisu 93. (Translated from German originals from 1908 and 1917.) Tampere: Vastapaino, 1999. Print.

Ukkonen, Taina. *Menneisyyden tulkinta kertomalla. Muistelupuhe oman historian ja kokemuskertomusten tuottamisprosessina.* Helsinki: Suomalaisen Kirjallisuuden Seura, 2000. Print.

—. "Muistitieto tutkimuksen kohteena ja aineistona." *Elore* 7.2 (2000). The Finnish Folklore Society. Web. 27 Feb. 2010. http://cc.joensuu.fi/~loristi/2_00/ukk200.html.

Östman, Sari. *Avaudutaan arjesta. Suomalaisnaisten Internet-elämäjulkaiseminen ja maskuliinisuuden esittämisen keinot.* Pori: Turun yliopisto; Kulttuurituotannon ja maisemantutkimuksen laitos; Kulttuuriperinnön tutkimus, 2007. Master's thesis. University of Turku. Print and Web. 27 Feb. 2010. http://smostm.files.wordpress.com/2007/11/avaudutaan-arjesta.pdf.

—. "Elämäjulkaiseminen – omaelämäkerrallisten traditioiden kuopus." Elore 15.2 (2008). The Finnish Folklore Society. Web. 27 Feb. 2010. http://www.elore.fi./arkisto/2_08/ost2_08.pdf.

—. "What is a Good Life Publisher Made of – and Why?" The International Journal of Interdisciplinary Social Sciences 5.5 (2010). 93-104. Print.

Sari Östman is a Researcher of Digital Culture at the University of Turku. During the last few years, she has held various positions in Digital Culture. She is interested in human/technology relationships and is currently working on her doctoral thesis about internet life-publishing.

Exploring Digital Identity: Beyond the Private Public Paradox

Stacey M. Koosel

PhD Student
Estonian Academy of Arts
Tartu mnt 1
10145 Tallinn
Estonia
stacey@artun.ee

1. Introduction: Defining digital identity

As new media is transforming culture, we transform ourselves into digital identities in the information age. Digital identities are who we say we are, when we are online. They can be a subtype of a public persona, an extension of our 'true' selves, or they can be completely fabricated and fantastical, to function as a mask to hide the identity of an Internet user from rest of the world. A digital identity can spin intricate, interconnected webs utilising creative, social and interactive platforms that enable them to share and perform to an open or closed audience (Cubitt). Both online identities and online communities are part of a virtual reality; simply put, a reality or existence that in most cases will only exist on the Internet and not 'offline' in real life.

The phenomenon of the digital identity has been referred to by many different terms including: online identity, online personality, digiSelf, virtual identity, avatar and online persona. These terms all refer roughly to the same idea, of an individual using a computer and creating a new identity for themselves on the Internet. It is important to clarify that the online identity is not a computer user in the traditional sense of man operating a machine, nor does it refer to any significance in the human-machine interaction. The significant interaction of digital identities are human to human or more specifically online identities interacting with other online identities in a virtual environment. The computer and the Internet are merely the medium, the software on the websites provides a platform or stage for online identities to perform or exist (Goffman). Digital identities are the content of the Internet; they are the performers who draw in the audience, and inspire the passive audience to become more active, interactive and creative. Online existence in online communities and digital identities are merely web-mediated human interactions. With this new medium comes new cultures, new ways of presenting the self and interacting with others as well as interacting with the medium itself. Sometimes we represent ourselves online in narrative form, other times we can use visual images, videos, audio and music. There is no limit to how or within what parameters we can express or present ourselves online, as Federman explored the idea through the metaphor of digiSelves, each with their own autonomous existence.

> What was once integral – our self, our person, our identity – is now split among our self in the physical world and our many digiSelves, each having an autonomous life of its own. Thus, we disconnect from the normal experience of physical and corporeal time and space when we live vicariously through our digiSelf on the Internet. This disconnection is significant and profound, as our consciousness becomes disconnected from our sensorium, extends in a real sense into the world's electronic nervous system and thereby creates the unique experience of separating our identity, or self, from our body. (Federman)

An individual can use an online identity to act as an extension of their offline identity, or they can use an online identity to mask and alter their offline identity and become someone or something else. This power of deciding what information to share and what information to withhold is a meaningful characteristic that the creator of a digital identity has online, and not in the real world. In real life it may prove more challenging to convince people that you are a different gender, age or species than it is on the Internet. The mask that the Internet user wears, can often be as simple as a pseudonym or 'handle' instead of their real name. This new or hidden identity can act as a mask to protect the privacy of the user or simply be an act of self-expression and entertainment. Derrick de Kerckhove, the former director of the McLuhan Program in Culture and Technology at the University of Toronto, predicted in 1995 that in the future "Changing our personal identity will become a primary entertainment, like a cosmetic surgery of the psyche" (de Kerckhove 20). This prediction is particularly interesting, because not only does it foretell the Web 2.0 revolution of user created Internet content, but all the creative platforms on the Internet that invite people to 'create a new user'.

Identities that are created online simultaneously flaunt and hide information, leading to a private and public paradox in their online existence. This contradictory nature characterises digital identity and their carefully protected private and public spheres of existence that one is caught between when creating an online self. It is possible that this drive to experience informational self determination, by creating a digital identity, is an important aspect of modern life. The creation of a digital identity and the paradox of wanting privacy and publicity at the same time, is perhaps only a microcosm of the larger informational trends in society and culture today. With the rise of the surveillance or transparent society (Brin), citizens of democratic nations have experienced a dramatic shift in how easily rights regarding privacy can be taken away. Voyeurism has become not only a recurring theme in entertainment but social policy in the information age. The latest controvery regarding personal privacy violations are the x-ray scanners and increaringly 'intimate' physical searches being used in airports where air travelers continue to be treated as criminal suspects (Paul).

The world of interactive multi-media has been in rapid transformation in the last decade. The dawn of personal computers and the accessibility of the Internet has changed the way people interact with media and with each other. The Internet has provided new interactive platforms of communication that were not available before, with traditional forms of media such as television, radio and print-journalism. Marshall McLuhan, a communications theorist, philosopher and media guru, prophesised his visions of the future long before the world wide web existed. In McLuhan's iconic 1962 media analysis, *The Gutenberg Galaxy: The Making of Typographic Man* he introduced the idea of the 'global village'. The term global village described a phenomena in which the world would become more closely interconnected like a village, and where the movement of information would instantaneously transmit from one point of the earth to another. McLuhan predicted the global village would happen with the rise of electric technology and mass-media. The global village, is now used as a metaphor for the Internet, as the movement of information has become instantaneous, connecting the world and people to an intense degree that was not possible before. McLuhan predicted that this change in culture and comunication would have far reaching and dangerous repurcusions, including the shift from literate man and culture to tribal man and a culture in which privacy does not exist.

The virtual environment fulfills social and entertainment needs by providing the Internet user with a sense of self and community. The more active user may create a website or write in an online journal (blog), or they may choose to only share pictures or videos. More passive forms of digital identity expression may only comment on things others have created online. In creating a digital identity there is the self-aware need to be seen or heard, or simply to participate and be a part of something bigger, such as an online community. However in the paradoxical nature of digital identity existence, sharing their real identity and information is often seen as potentially dangerous. This creates a situation in which the Internet user is simultaneously seeking publicity and privacy. The need for attention, the need to share and be a contributing part of the online community, must obviously at times outweigh the need for privacy. Creating an artificial identity can liberate the Internet user to provide or share select information, without feeling exposed or like their privacy has been invaded. The privacy versus publicity aspect of digital identities can often lead to a very interesting paradox in which the digital identity may want to be public or even seek fame yet has a strong desire to guard its privacy and real life identity at the same time. We are who we say we are online; any identity can be assumed, used or discarded. It can be suggested that online personalities will be greatly influenced by the online communities with which they associated themselves. Like a chameleon that changes colours to suit its background or settings, so the online personality can change to adapt and fit in to an online community, or in some cases to create a community of its own. A digital identity is a manifestation of a real life identity that exists on the world wide web, but it need not be a single identity. An individual can have numerous different digital identities to serve different purposes and needs. For example, on some websites a person may choose to use his or her real information, a second website might be slightly modified to thinly veil the identity, and a third digital identity could be a complete work of fiction.

2. Transcending mediums

As McLuhan prophesised the current reality of the 'global village', an online identity can be likened to be the global village's citizen or 'netizen'. The birth of a digital identity can start as easily and simply as creating a name, account or handle to register on an Internet website, and can be as elaborate as an online existence that spans over many different websites, including a multi-media trail that can include anything from photographs, text, videos, music and even live webcams. Some Internet users engage in electronic exhibitionism and strive to attract as much attention as possible, and become celebrities by the careful construction of an online identity. The real life identity of a digital identity can be kept completely separate and anonymous, even if the particular digital identity is well known on the Internet (Internet famous).

This selling of the 'real life self' or 'real life stories' online can be seen as the commodification of the individual. Selling what would otherwise be private human existence to the public as a form voyeuristic entertainment has been a popular subject of reality TV, websites and weblogs. Traditional celebrities or 'stars' were made and marketed by businesses and institutions that had a vested financial or collective interest in making a particular individual well known to the public. However in the blogging era, an entire marketing team or appearances on traditional forms of media such as television, radio and newsprint are not needed to make a person on the Internet well known and famous enough to transcend mediums, from the virtual to the material. It is possible that a digital identity is not necessarily being commodified or being made into an object, as they are existing or performing to the Internet community of their own free will. However, the key difference between a human and objects when being sold as commodities is that a human is able to try to "keep the magic alive and avoid being discarded. This an object cannot do." (Hillis 352)

Transcending mediums is a term that I use to describes the phenomena when a digital identity whose existence was created on the Internet reaches a critical mass of popularity or potential commercial sucessess on the Internet and is offered to continue their work in a more traditional and respected form of media, such as print journalism or television and with the change of medium comes a change of identity. For example, a blogger who has a very large dedicated fan base may be offered a book deal and become a best selling author. This is a goal for some bloggers, as it brings much more financial gain as well as 'real world fame'. The very important detail in the scenario that supports the theory of the private and public paradox of digital identites is the way authors will present themselves in traditional versus virtual media. On the blog of the very popular author, his or her name is nowhere – the blog is written anonymously or under a pseudonym. When this blogger transcends mediums and becomes a best selling author, suddenly his or her name is on the dust jacket of the book, in print for all to see in newspapers and interviews on television. The blog readers finally discover the 'true identity' of the author 'behind' the blog, for most of the blog readers it will be the first time they can put a name and face (author) to the content they had been enjoying for years. What is it about the Internet that is deemed dangerous, untrustworthy, unestablished and not real – that the readers of a book or newspaper may be deemed a worthy enough audience to know the real name and face of the blogger but not the blog readers? Is it the perceived scale or value of the audience? Or is it the nature of the online enviroment and digital culture itself? The exciting and interesting side product of studying digital identity is that in studying digital identity, we study the cultural artefacts that give us insight into the culture in which the creator belongs. To study digital identity as a cultural artefact and to be able to speculate on what the culture and enviroment is around the creation of this identity has a double reaching effect. Firstly a greater understanding and knowledge of digital culture, and seconldy the next layer of understanding modern culture and the world around us. Studying digital identity is the key to understanding the values, stigmas and characteristics of the modern world.

When a digital identity transcends mediums, from the relatively new and unknown world of the Internet fame to the established, respectable, traditional form of media such as print journalism (newspapers and magazines) or printed publications (books) they have arrived as a real, published author who can share with the public their true identity. Perhaps it is the equal opportunity of the Internet that enables anyone to have a website or a blog, which degrades the status of blog writers to be anonymous or very loosely connected to their true offline identity. The fact that one medium (the Internet) settles for a pseudo or fabricated identity and another medium (published print) requires a real name and identity is a very interesting clue to why digital identity exists in the first place. "Environments are invisible. Their ground rules, pervasive structure, and overall patterns elude easy perception." (McLuhan) It is possible that it is the medium itself that dictates how we present ourselves, the invisble yet socially accepted rules of the electronic enviroment in the case of digital identity.

3. Informational self determination

The information age gave people new ways to fulfill their basic human needs, to be connected to others, self-expression and to belong to a greater community. Interactive technology provided new dimensions of existence, shaping the new globalised world into a smaller accessible global village. Web 2.0, the creation of web platforms that enabled Internet users to carve out their own little pieces of the Internet and create their own terms of existence has set the stage for the creation of the digital identity. The important aspect of the digital identity that Web 2.0 platforms like Flickr and Facebook understand, is that digital identities want the right to informational self-determination. Though it is arguable how much power the digital identity truly has over their personal information on Web 2.0 platforms and how much is done to give the illusion of having control of their own information. Digital identities are created when users are given the right to decide what information is public and what information is private. This is acknowledged by the majority of Web 2.0 platforms, which offer a variety of privacy settings.

Informational self-determination can be defined as an individual's right to decide what information should be communicated to others and under what circumstances (Westin). This term was originally used to outline a new law structure that would give people the right to protect their own personal information in real life. With the relatively new existence of digital entities, with real life traits and information that may or may not exist in real life as well, the idea of informational self-determination is brought to a different level. Digital identities have more informational self-determination than the individual who created that digital identity has 'offline' in the real world. This freedom of choice, to be selective or manipulative with information regarding the self, benchmarks a new era in human existence. A good illustrated example of digital identities and their informational self-determination was published on July 5th 1993 in *The New Yorker* magazine. A cartoon shows two dogs talking, one of them sitting behind a computer with his paws on the keyboard with the other dog sitting on the floor. The first dog is telling the second dog: "On the Internet, nobody knows you're a dog." The recent update to the joke, and social comment on the current surveillance society we live in, is: "...but they know you buy dog food."

4. Surveillance society, transparent society

While concealing one's identity might be easy to do online, it is a feeble defense against the predations of government and corporate power in real life. Though the digital identity might go to great extremes to protect their real identity when they are online, the real world they live in is becoming a surveillance society. With the belief that surveillance is synonymous with security, the world could enter an Orwellian dystopian existence overnight without any large public outcry or mourning for any loss of former civil liberties. The appetite for information and willingness to be monitored seems to characterise the modern world as an information society. Reality television is a popular form of voyeuristic entertainment, surveillance cameras do not bother citizens and many computers come with webcams as standard features, which begs the question: is the need for privacy history?

The rise of the surveillance society, would seem like a science fiction depiction of the future; however the surveillance society already arrived some time ago. The biggest Western surveillance society to date is the United Kingdom, with 4.2 million surveillance cameras (CCTV) installed and functioning in 2006, which is equal to one camera per fourteen people (BBC News). There are of course, polar opposite views on whether the surveillance society is progressive or regressive. Many civil rights activists fear that the power and authority can be misused, and can take away civil liberties formally enjoyed by citizens of democraticies.

However not all interpretations of the surveillance society are negative, some theorists have optimistic views on the surveillance society. David Brin, the author of *The Transparent Society*, believes the rise of the surveillance society is imminent and unavoidable, however the repercussions need not be negative if the transparency is reciprocal. Brin believes in "a future where privacy and anonymity have been rendered obsolete by technological change" (Stenz). The opposite of a transparent or open society is a secretive society that protects not only citizens but the larger powers, such as world governments, businesses and the criminal underworld.

Marshall McLuhan believed that as man becomes more interconnected there will be a shift from the individualistic literate man back to the collective tribal man. "Privacy, like individualism, is unknown in tribal societies, a fact that Westerners need to keep in mind when estimating the attractions of our way of life to nonliterate people." (McLuhan 129) Tribal man, according to McLuhan has no need for privacy, as the need for community and the global village would come before the need for individuality and fragmented self that characterised the literate man. Although tribal man may exist in the information age, where the need for privacy would be subversive to the values of the surveillance society – digital identities are collectively identified by their need for privacy and information self-determination, which would classify them as literate man, despite their tribal urges and need for community. The private and public paradox of the digital identity is something that complicates and contradicts basic user gratification theories and traditional understandings of how audience and media interact. If media is an extension of man, then digital identities are an extension of the human psyche and other realms of existence, not merely digital copies of our real life selves. The digital identity's paradox of wanting privacy and publicity at the same time is a microcosm of the larger trends in society and culture today. With the rise of surveillance or transparent society, the very definition of privacy is changing for citizens of the free world.

The digital identity, unlike the Internet user, can easily enjoy a state of assumed identity or anonymity – in many ways the Internet user has more freedom on the Internet than they do in real life. Perhaps that is what has made the Internet such a powerful creative platform, where people from all over the world can come together to create new communities and new identities; it is a brave new world. This is a grassroots movement with major cultural and social repercussions that have not yet been understood. Perhaps a motivational force for an individual to create a public yet private entity and exist on another level as a digital identity, is driven by the need to leave a mark behind. "The only thing we know about the future is that we die. The brief, messianic experience of the cybernetic supernatural is an attempt to control that future by creating a bubble of eternity in the fluid materiality of time." (Cubitt 18) Informational self-determination is something that we have on the Internet and not in the real world. The idea of a fractured self or post-modern definition of identity easily lends itself to the idea that who we are in real life does not have to be congruent with who we are online. (Boardman) We become more fluid and ephemeral online, able to express ourselves to closed or open audiences (Cubitt) in ways we simply physically could not in the real world.

5. Conclusion: Future research and interdiscplinary studies

Digital identities are the forms that humans assume to navigate the virtual world; they are the citizens of the global village, the content of social media. A very complex relationship has developed in the link between the online self and the offline self. The paradox of wanting privacy and publicity at the same time is only one of the contradicting yet defining characteristic of online existence. The use of pseudonyms, assumed and multiple identities can simultaneously distance the user from their online persona, yet at the same time provide the user a more authentic representation of themselves than they could express in the real world. As more and more people have access to the Internet, the population and popularity of digital identities will continue to grow exponentially. Web 2.0 platforms enable Internet users with limited technical skills and computer competence to create their own websites with the push of a button. The popular blogging website 'Blogger', refers to itself as 'push button blogging'. The popularity and easy accessibility of these user-created content-based websites gives Internet users the opportunity to share ideas, photographs, stories and perhaps most importantly, themselves. To share information about yourself on the Internet is to create information about yourself and to struggle with the private and public paradox of creating and maintaining a digital identity. There are many questions and themes of digital culture and identity that can be explored and analysed utilising a variety of frameworks, research methods and methodologies. What information is safe to share with others on the Internet? Should a digital identity be based on truth or fiction? Is there real world stigma against virtual reality based creations of identity? Is there stigma attached to fictional identities on the Internet? To what degree is the creation of a digital identity an act of self-expression or artistic expression? Or, are all digital identities conforming and adhering to an unspoken yet understood formula of tribal values?

The creation and utilisation of a digital self to function in the online community is a relatively new media phenomenon that requires the creation of new paradigms. Traditionally, media or audience studies and methodologies are often insufficient when studying the link between the user and their digital identity. The information revolution has spawned a new type of audience, no longer passive and consuming but instead active and producing (Gauntlett). Creating a digital identity is potentially a creative and artistic act and can be a fulfilling form or self-expression for some Internet users. Artistic and creative research methods provide new ways to study and understand the relationships that an online user has with their digital identity or avatar. Traditional methods of media research and audience studies frequently use questionnaires or surveys. However when dealing with a topic as encoded and complex as identity and representing identity online the use of traditional research methods can be problematic. It is also possible that there is social stigma about the new online culture and post-modern ideas of identity that do not translate to popular and modernist philosophies of a singular, unfragmented self such as the idea of 'one true self'.

It is possible that the evasiveness or undercurrents of hostility when discussing the topic of digital identity when conducting a study in a traditional method such as interviews and questionnaires is firmly rooted in the private and public paradox of digital identities. As the Internet and Web 2.0 platforms create an environment where people can express themselves is a much more liberated and less consequential space than in the real world, discussing what they do on the Internet and what kind of identities they have created may feel like an invasion of privacy itself.

In my forthcoming doctoral thesis I will explore different methodologies for studying digital identity and compare the findings of different methods used to analyse and explore the idea of the online self. The exploration of methodology will also outline the paradigm and assumptions behind each method used, as well as explore problems and discourse. Currently the topic of digital identity is being researched using a variety of different methods including (but no limited to): digital ethnography (Michael Wesch), cyber psychology (Andra Siibak), digital narratives (Ruth Page), creative research methods (David Gauntlett) and visual ethnography (Sarah Pink).

The idea of social media such as weblogs (blogs), Facebook updates or Twitter tweets representing a modern form of narrative is very exciting. It would be interesting to find out if the content of social media can be interpreted as modern narratives, if then digital identity can be viewed as a modern storyteller or folklorist. Though it is also possible to view digital identity as the story themselves, as they are the creative expression of an offline user to function in the online world, almost like a translation or metamorphosis of the self. Although the virtual environment has provided new modes of existence in the form of new ways to present ourselves and communicate with others, there are still very traditional aspects to new technology, they still exist to fill cultural and basic social needs. A seemingly basic and global human need is that of storytelling and storytellers, which has evolved to a more diffused cultural background and larger anonymous audience in the information age. "Are there pan-cultural forms defining a generic human form of life? This does not seem a far-fetched hypothesis, but the issue needs to be settled by wider comparative studies. What is true is that every culture of which we know has been a storytelling culture." (Brockmeier and Harré 266)

The direct translation of narrative studies and sociolinguistics to the online interactive environment is a challenging task as it requires new paradigms of stories and storytellers, self and community and other terms which have had to take on new meanings in a virtual environment. Another complication about studying digital identity as an online narrative and utilising literate and oral model resources (Page 2010) is that the online environment can at times be a purely visual environment, which is sometimes not only the context but the 'text' itself. To only pay attention to text in an online environment is to lose sight of the larger picture, in the sense that studying the online environment by only analysing texts and narrative structure would be ignoring a largely visual environment. However utilising a variety of methods to study digital identity and the digital environment as both visual and a narrative can perhaps reveal a lot more about the phenomenon than simply concentrating on one or the other. In this sense, virtual ethnography and visual ethnography are both very interesting methodologies to use when studying digital identity.

In my own research I am open to exploring multi-disciplinary approaches and comparative studies. To probe the subject of digital identity, includes visual ethnographic analyses of the images and creative research methods (Gauntlett) to explore the motives of the offline Internet user for creating content and a sense of 'self' online. The narrative structure of online communication would be interesting to delve into, as there is a separate unique vernacular that has developed on the web. Much like William Labov studied the African American English vernacular and in doing so developed the methodology for sociolinguistics, so studying the vernacular used in online communities could help develop a framework and methodology for studying digital ethnography and digital culture.

The work of Erving Goffman has been essential in the study of digital identity, particularly "The Presentation of Self in Everyday Life". Goffman's idea that social interaction and face to face interaction are framed as theatrical performances, with the metaphor of the self as a performer and life being a stage with the back stage symbolising privacy and private lives and the front stage symbolising the public and public actions, easily translates to the notion of the digital identity being a mode of performance or theatrics that utilises the online environment as a stage on which to perform. Georgakopoulou further expands Goffmans exploration of self as performer, story telling and propagandists: "As Schiffrin (1990) has explicated it, drawing on Goffman, storytellers can present themselves in the capacity of (a) animator (the aspect of self which physically produces talk), (b) author (the aspect of self responsible for the content of talk), (c) figure, the main character in the story, someone who belongs to the world that is spoken about and not the world in which the speaking occurs, and, finally, (d) principal, the self established by what is said, committed to what is said.... Through such manipulations of their kaleidoscope of selves, storytellers can diffuse their agency or responsibility in the social field, create a widened base of support for their views and beliefs, or, generally, cast positive light on them (e.g. see Hill 1995)." (Georgakopoulou 16) Much in the same way a digital identity can also "manipulate a kaleidoscope of selves" (ibid.) to multi-function and fit into different environments and roles that have been outlined online. As the study of digital identity and digital ethnography are still in their relative infancy compared to more established fields of cultural studies such as narratology, film and media studies, social anthropology, visual ethnography, etc., it is up to the researchers and theorists in the field of digital ethnography today to develop new frameworks to base theories on and create new methodologies to probe a new form of culture and identity. The fact that the Internet is such a diverse, chaotic, rapidly changing organism makes it very hard to pin down and analyse or generalise about a particular topic or phenomenon. Cultures are always changing and evolving, but on the Internet the concept of time and space is much more asynchronous and fragmented than in other traditional forms of media and communication (Page). It is an exciting idea that the Internet is not simply a new form of technology, a communication tool that we utilise much like any other form of media, but rather it is the 'digitisation of us' and that we become on some level the content of the Internet, what is commonly referred to as Web 2.0 (user created Internet content). "…the key cultural consideration of the Internet is not so much the digitization of information, but the digitization of us." (Federman) The interesting question concerning digital identity can then be asked, to what extent is Web 2.0 user created Internet content and not content created Internet users?

References

BBC News. "Britain is 'Surveillance Society'." 2 Nov. 2006. Web. 15 Jan. 2010. http://news.bbc.co.uk/2/hi/uk/6108496.stm.

Boardman, Mark. *The Language of Websites.* New York: Routledge, 2005. Print.

Brockmeier, Jens, and Rom Harré. "Narrative: Problems and Promises of an Alternative Paradigm." *Research on Language and Social Interaction* 30.4: 263-283. 1997. Print.

Brin, David. *The Transparent Society: Freedom vs. Privacy in a City of Glass Houses.* New York: Basic Books, 1998. Print.

Cubitt, Sean. *Digital Aesthetics.* London: Sage Publication, 1998. Print.

De Kerckhove, Derrick. *The Skin of Culture: Investigating the New Electronic Reality.* Ed. Christopher Dewdney. London: Kogan Page, 1997. Print.

Federman, Mark. "McLuhan Thinking: Integral Awareness in the Connected Society." *The McLuhan Program in Culture and Technology.* 10 Feb. 2003. Web. 29 Dec. 2011. http://individual.utoronto.ca/markfederman/IntegralAwarenessintheConnectedSociety.pdf.

Gauntlett, David. *Creative Explorations: New Approaches to Identities and Audiences.* New York: Routledge, 2007. Print.

Georgakopoulou, Alexandra. *Small Stories, Interactions and Identities.* Amsterdam: John Benjamins, 2007. Print.

Goffman, Erving. *The Presentation of Self in Everyday Life*. New York: Anchor Books, 1959. Print.

Hillis, Ken. *Digital Sensations: Space, Identity, and Embodiment in Virtual Reality*. Minneapolis, MN: University of Minnesota Press, 1999. Print.

McLuhan, Marshall. *The Gutenberg Galaxy: The Making of Typographic Man*. Toronto: University of Toronto Press, 1962. Print.

McLuhan, Marshall. *Understanding Media: The Extensions of Man*. Toronto: University of Toronto Press, 1964. Print.

Page, Ruth. *New Perspectives on Narrative and Multimodality*. London; New York: Routledge, 2009. Print.

Paul, Ron. "Are Air Travelers Criminal Suspects?" *Homeland Stupidity*. 22 Nov. 2010. Web. 29 Nov. 2010. http://www.homelandstupidity.us/2010/11/22/are-air-travelers-criminal-suspects.

Pink, Sarah. *Doing Visual Ethnography: Images, Media and Representation in Research*. London: Sage, 2006. Print.

Siibak, Andra. "Constructing the Self through the Photo Selection – Visual Impression Management on Social Networking Websites." *Cyberpsychology: Journal of Psychosocial Research on Cyberspace*, 2009. Web. 20 Nov. 2010. http://www.cyberpsychology.eu/view.php?cisloclanku=2009061501&article=(search%20in%20Issues).

Steiner, Peter. *The New Yorker*. 69.20: 61. 5 July 1993. Print.

Stentz, Zack. "Unmasked Society." *Metroactive*. 6 Feb. 1997. Web. 10 Jan. 2010 http://www.metroactive.com/papers/metro/02.06.97/cover/brin1-9706.html.

Wesch, Michael. "From Knowledgable to Knowledge-able: Learning in New Media Environments." *Academic Commons*. 7 Jan. 2009. Web. 22 Nov. 2010. http://www.academiccommons.org/commons/essay/knowledgable-knowledge-able.

Westin, Alan F. *Privacy and Freedom*. New York: Atheneum, 1967. Print.

Stacey Koosel is a Canadian-Estonian PhD Student at the Estonian Academy of Arts in Tallinn, Estonia. Her research and lectures are in the field of social media, digital identity and media ecology. Her current projects include: an ethnographic study of Facebook users, her doctoral thesis on Research Methods in Digital Culture and a book on Digital Identity Narratives.

From Landscape to Multi-layer Landscape: Landscape as a Tourism Resource on Web 2.0

David Casado-Neira

Professor
Faculty of Education, University of Vigo
Avd. Castelao s/n
32004 Ourense
Spain
dcneira@uvigo.es

1. Landscape as an aesthetic experience

Landscape is a very complex phenomenon. From a geomorphological view, landscape is the result of years of progressive and deep changes. The forces of nature are sometimes abrupt and unexpected, but the majority of the geomorphological transformations are too slow to be assessed with a simple glance. From historical, economic, cultural and political views landscape can also be explored scientifically as a transformation of a territory, which is a consequence of human activity. In addition to the complexity of actors and factors behind the emergence of a landscape, the ambiguity of the concept of landscape itself may also be taken into consideration. Shunsaku Miyagi explains that the Japanese ideograms for the English word 'landscape' are 'wind' and 'view', *fukei,* which are related to the ideas of space and perspective, or seeing from a distance. Furthermore, Miyagi points out that: "Implicit in the word *fukei* is a recognition that the landscape is something over which we human beings do not exercise complete control, a perception that the landscape involves phenomena without clear, distinct forms" (4). Under this consideration landscape seems to be more the result of an infinity of forces working in a plastic material, partially hidden, unclear and even out of human control. Landscape emerges as an aesthetic experience, a product composed of several layers. What we appreciate as landscape is a view – a excerpt – of different parts that have some meaning for us (Simmel 270; Turri 218). Landscape could be initially defined as "all the visible features of an area of countryside or land, often considered in terms of their aesthetic appeal" (Oxford Dictionaries). Thus we can talk about three types of landscape: a beautiful landscape (e.g. a birch wood), a terrifying landscape (e.g. a run-down city), and a spoiled landscape (e.g. a dried up lake), keeping in mind that there is some kind of ideal image being used as rule for comparison that leads to aesthetic canons given by social and cultural values. In other words, we could say that only what has value is perceived. Furthermore landscape is a pure aesthetic experience, built as "the recurrent reference to a vocabulary of views and panoramas, to an apparently pictorial mode of looking" (Green 37). Landscape is not neutral but is a social construction, a way of viewing and experiencing a territory that acquires sense to the person by using some artistic patterns socially given (Roger).

Questions like how a piece of land, water (sea, lake) or sky gains an aesthetic value, or how these aesthetic patterns acquired social value, for example in the construction of national identities, for specific communities or as idealised products, are interwoven with the powerful forces that exist between social actors (between state and citizens, between community members, or between market and consumer). Landscape is a way of seeing, a way of recreating an environment. If the way we see is socially given, landscape is a political question (Foros do Instituto de Estudos das Identidades). Thus the above definition of landscape could be redefined in the following way: "all the visible features of an area of *territory*, *always* considered in terms of their *constructed* aesthetic appeal".

2. Emergence of the multi-landscape on the Web

Images of landscape are one of the main resources used in the touristic promotion of a place. Tourism is an experience with a deep aesthetic component, and, especially in outdoor tourism, landscapes play a central role. Many outdoors activities, such as hiking, skiing, hang gliding, etc., are ways of enjoying landscapes: the view can be the reward or, at least, a very important part of the experience, as in the origin of the touristic landscape experience itself. In 1336 the Renaissance poet Francesco Petrarca climbed Mont Ventoux and wrote: "I pray you, to reviewing the experiences of my entire day. At first, owing to the unaccustomed quality of the air and the effect of the great sweep of view spread out before me, I stood like one dazed. I beheld the clouds under our feet, and what I had read of Athos and Olympus seemed less incredible as I myself witnessed the same things from a mountain of less fame. I turned my eyes toward Italy, whither my heart most incline" (Petrarca and Robinson 314-315).

A view can be shared through a letter to a father (as in the case of Petrarca), or through an exhibited painting or a digital photograph. Today tourism is a widespread phenomenon. In recent years we have seen how the Internet is changing the rules of the tourist market. It has brought not only better access to information, increased possibilities of information sharing and new ways of consumption. New media, specifically digital photography, has become a new tool with transcendental impact (Lister). The birth of Web 2.0 (O'Reilly) changed the ways tourist destinations are represented and how landscape is recreated. Sharing information about destinations and exchanging photographs between visitors in online services (like Panoramio, Flickr, Spottedbylocals, Qype, and others, plus internet forums or private blogs) are transforming the sources of information and the way destinations are represented.

Institutions and tourist services still provide the basic representation of destinations, and marketing campaigns are the most important tool with which to create and transmit a landscape as a consumer product. A picture of a landscape for touristic promotion tells us something not only about the environment where the photograph was taken, but also about what value the landscape has for the promoters of tourism. However, visitors also add layers of information to these institutional or promotional images of destinations. The shift from a physical heritage to a digital heritage represents more than the virtualisation of environments: it is their re-representation. Online access to information makes possible, on the other hand, access to no-presence tourism in a more intensive way, similar to what diaporamas, slide-shows, travel documentary films or travel guides and coffee table books offer. On the other hand, this supports the birth of new representations of tourist destinations based on vivid, personal experience. Conventional images of a landscape are no longer the only available representations of a destination. The sense of a landscape would then no longer be the result of fixed conventions that make a landscape into an iconic figure, but rather would be the consequence of the actions of either individual, or various, photographers, mainly amateurs. Does this mean the democratisation of the representations of an environment, or the emergence of a new social consensus on landscape? More than a new aesthetic consensus we may be assisting in the creation of an intensification of meanings and senses of the previous aesthetic patterns.

Multi-representation may be a phenomenon given not only by the sharing possibilities of Web 2.0. In contemporary societies indetermination, vagueness, inconsistent or weak hierarchies are the main characteristics of a new *Zeitgeist*, characteristics also present in other levels of social life where the identities in the liquid modernity emerge (Bauman) or in primary elements of human constitution, such as embodiment processes. "In Western cities the actor creates the knowledge he has of his own body in every-day life through a jumble of more or less similar heterogeneous models, regardless of the compatibility of the borrowings. The current profusion of body images is an evocation of the dismembered body of the schizophrenic. The actor rarely has a coherent image of his body; he transforms it into a tissue, streaked with different references. There is no theory of the body generally assumed. The person, in having to choose in a nebulosity of possible knowledges, swings between them without finding one that fits him entirely. [My own translation]" (Le Breton 91). By replacing the word body with the word landscape the text would still preserve the same underlying sense. Body or landscape, experiences of embodiment or aesthetic experiences of the environment, seem not to be so far away from each other (Cloke and Jones; De Klerk; Rose). This leads to the principe of two possible ways of being-in-the-world (Varela, Thompson and Rosch), one by living a to-inside experience (by realising the physical self), another by living a to-outside experience (by realising the physical environment) directly related to the idea of landscape. Arnold Berleant has expressed it as a phenomenological question: "The consciousness of self, of the lived-body and of lived-space must be complemented by a recognition of environment as exerting influences on the body, as contributing to the shaping of the body's spatial sense and mobility, and ultimately to the definition of its lived-space" (119). The aesthetic experience of environment would be considered a definition of landscape itself.

In the same way, shared landscapes are excerpts of the personal experience of the photographer, and Web 2.0 opens the door to wide, dynamic and self-selected representations of the environment. Let us choose a paradigmatic touristic destination like the Maldive Islands as a pure example of a tropical tourist destination. Viewing the uploaded photographs on photo sharing services like Flickr (www.flickr.com) or Panoramio (www.panoramio.com) hundred of pictures are found, many of them with the iconic elements of a paradisiacal destination: sea, palm trees and sand beaches with no trace of human presence. Despite the fact that these photographs are non-commercial and have no promotional purpose, they represent the common image of a paradise island, a very common image in touristic advertisement. These photographs could be used for promotional purposes; they transmit the 'feeling' of the Maldives as a tourist destination. It is also possible to add many visual layers and more meanings, which would sometimes be in harmony with the iconic image of a paradise destination; others may be complementary, or even contradictory. The next photographs (see Photographs 1 and 2) are also landscape pictures. However, would it be possible to say that they are common in tourist promotions?

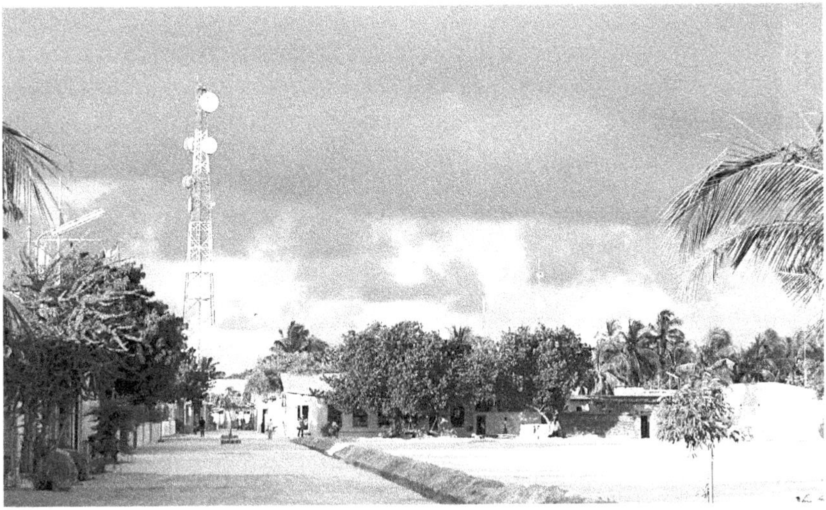

Photograph 1: Maafushi (Dhivehi) is one of the inhabited islands of Kaafu Atoll. By Nattu from Malé, Maldives (Creative Commons Attribution 2.0 Generic license)

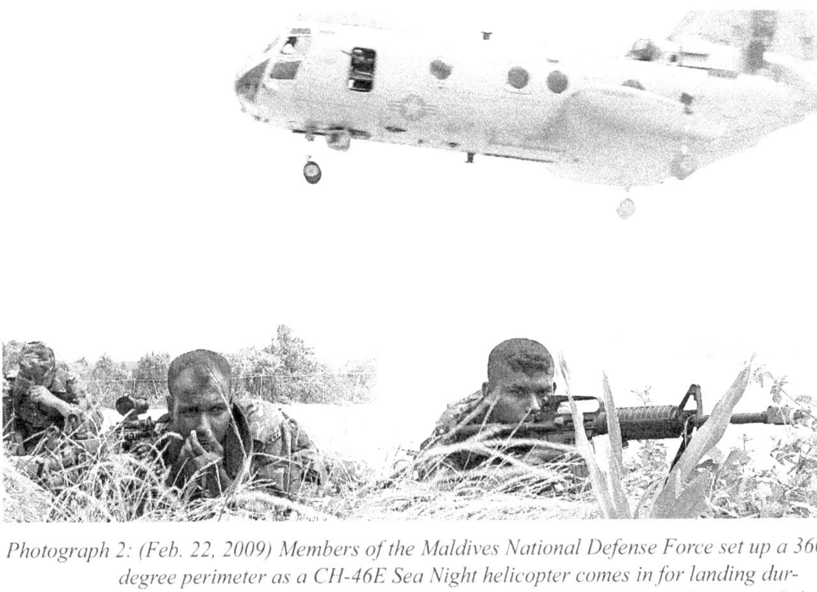

Photograph 2: (Feb. 22, 2009) Members of the Maldives National Defense Force set up a 360 degree perimeter as a CH-46E Sea Night helicopter comes in for landing during Exercise Coconut Grove. By U.S. Marine Corps, photograph by Lance Cpl. Jesse Leger (Public Domain)

A piece of information that would not be normally presented as interesting or even attractive for visitors is given. The values underlying the representations of environment (landscape) lead to different meanings, some of which will be used in institutional representations of environment (i.e. landscape for touristic promotion), for private meaning (the memory of a moment, photographer's self promotion, etc.), other purposes (Marine Corps publicity…). However, independently of the intention of the social actor (institutional, non-institutional), every view of the photographed reality gives us more and more information on the destination. Even when the purpose of the shared photographs wasn't to do so. Sometimes the same values and meanings are found in the similar images, while at other times the same values and meanings may find a visual representation through other icons ('what represents paradise – a palm tree or inhabitants at the local market?'), and at yet other times the same icons are linked to divergent values (the sea: viewed for contemplation or as a space for sort activities?); or different values and meanings become relevant (paradise or touristic exploitation?). Web 2.0 creates a multilayer representation of environments with interlinked values (see Figure 1).

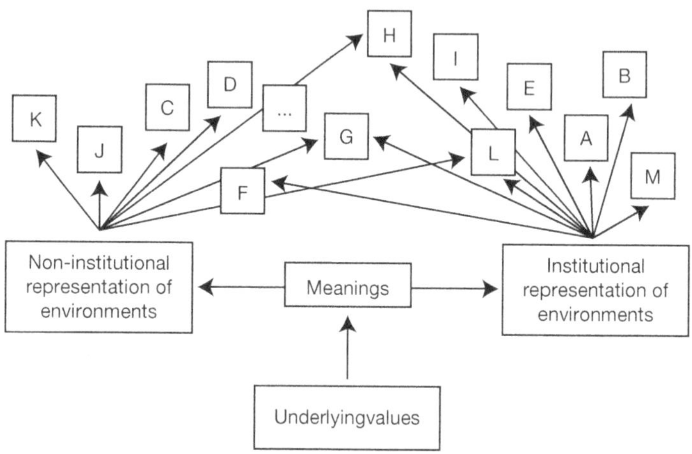

Figure 1: Image meaning for non-institutional and institutional representations of environment.

Some years ago photograph sharing was limited to small groups and audiences. Due to the explosion in possibilities of sharing information on the web, the repository of images from individuals is quickly growing, and with it the different meanings given to the elements chosen to be part of the landscape – even in cases where the underlying values were the same. Landscapes are becoming more complex, and richer in meaning and given information. The implicit values of the representation may correspond or vary between images, authors, institutional or personal views. My hypothesis is that the richer (more full of meaning) a place is, the more interesting it would be for a wider public. Every landscape on Web 2.0 is a multi-landscape. An environment can become a single landscape or many landscapes. In a single landscape only one voice is talking, while many voices construct a multi-landscape.

3. Applying concept of multi-landscapes to outdoor tourism

In March 2009, an organised group of hikers was asked to take three pictures of the nicest, the most unpleasant and the most typical scenes during some outdoor walks in Galicia (Northwest Spain). The group was composed of a total of 56 participants, both male and female, aging between 18-62 years. Each participant decided individually to upload the photographs to the given classification (nice, unpleasant or typical). The photographs were shared for a period of six months in networking site with public access created for this purpose. The aim was to create a situation in which participants were able to upload their photographs for sharing purposes. The goal of this project was to document how hikers perceive the landscape in order to research the touristic possibilities of this region.

The chosen destination was a protected area at the border between Portugal and Galicia. In fact this area is comprised of two natural parks on either side of the border (the Peneda-Gerês National Park with 702.9 km^2 on the Portuguese side, and the Baixa Limia-Serra do Xurés Natural Park on the Galician side with 297.6 km^2). The park is a highland area with massive granite outcrops, poor soil vegetation and with a small population who make a traditional living raising cattle. Granite (also used also for construction) and water have played an important role in the image and representation of this area, these two elements being the icons in the images of landscape.

For the analysis, every photograph was briefly described, tagged according to the items shown and according to the field size (long shot, full shot, medium shot). These photographs were compared, under the same theme and field size patterns, with those used in touristic promotion of the parks and secondary information materials (printed brochures, internet sites and local information stands and museums) by the two main institutions promoting this area as a touristic destination: Instituto de Conservação da Natureza in Portugal, and Turgalicia in Galicia.

After analysing the landscape photographs used by the tourism agencies for promotion, some patterns were found. There are three main elements: animals, villages, and mountains in the distance, which lead to three motifs.

- 'Animal' motif: A limited serial of animal portraits from the common fauna (like *aquila chrysaetos, vulpes vulpes, rana perez, lutra lutra...*) and photographs of cattle herds in open spaces are the third motive according to the number of cases.
- 'Architectonical' motif: These images pay major attention to views without human presence. If some trace of human presence or civilisation is offered, it is always with photographs of vernacular architecture without human figures. Elements of folk culture are mystificated as 'out of time' items, immutable witnesses to history, as they were not perturbed by social transformations they become like totems of ethnic/national identity. In this case photographs of mountain villages or photographs of single buildings (stone granaries, bread ovens, etc.) are the main motif.
- 'Mountains in the distance' motif: The most commonly present and important motif of all three. Due to the granitic nature of the ground and the poor soil vegetation, granite outcrops are the most characteristic element in landscape. Peneda is internationally known as an important rock-climbing destination. Curiously, no climbing mountains are shown, the mountains are always presented as scenery where only the distance plays a role. These images are presented like the paintings of Romanticism in which the forces of the nature can be felt. Granite plays the main role.

These are the main motif in the representation of the two parks; no significant differences were found between the Portuguese and the Galician landscapes. The same elements are limited and a cultural similarity can be found on both sides of the border, which may be attributed to a similar background for territory meaning-giving and interpretation by social actors.

The institutional landscape is a business card to visitors, but for us the question now is if the touristic promotion of the region is based on images and representations that are emotionally exciting for the visitors. The aim is not to detect what elements can be taken into consideration for future promotion campaigns, but to show that it is possible to create a multi-landscape with the same elements gaining meaning. Web 2.0 makes possible the diversification of representations and their spread to potential visitors further over the institutional promotions.

The group of hikers was asked to take some pictures at the natural park under the three given categories: nice, unpleasant and typical. The intention of using these categories was to detect which elements have a more positive and pleasant incidence (assertive icons), which have a negative connotation (no assertive icons), and which are considered the representation of the most characteristic (reinforcing icons). Some similarities and divergences were found between visitors' images and institutional ones. This experience shows that the way territory is perceived and represented (landscape) by the promoters of tourism wasn't the same as the way it was represented by the hikers. In some cases, hikers paid attention to similar elements to those used in touristic promotion, while in other cases they paid attention to different ones. Initially it was expected that the participants would reproduce the same aesthetic patterns present in the promotional pictures: beautiful landscapes and images based on common values (pure nature, views with skylines or far distances, vernacular architecture and elements associated with an ideal rural environment). It is true that some of the images could fulfill the pattern of the existing promotional images, even with a higher artistic intensity, but also that the majority of the others that are being introduced don't only contain new and complementary elements on the excerpts, but also offer different viewpoints.

- **Nice (assertive icons):** This is the most interesting category of photographs for promotional purposes. Under the category 'nice', people gather the pictures that best transmit the spirit of the place, which makes these photographs more likely to be shared with others. People like to show the pleasant (nice) and genuine (typical) aspects of the visited place as evidence of having been there, and these images are often used as common souvenirs in t-shirt and postcards. Under the category 'nice' we find conventional photographs of mountains in the distance, and other themes like panoramic views, vegetation and crops, or human presence in a group (no animals were photographed, perhaps due to the difficulty of taking this kind of picture when hiking). Mountains are not as they tend to be in tourism promotion, but the pictures reproduce the same schema, probably because of the impact of the media, national education and artistic canons determined by the persistence of the aesthetic patterns of Romanticism in contemporary societies when contemplating natural and/or rural environments. Photograph 3 offers an example of a conventional image similar to those used for promotional purposes in the use of natural elements (granite and water) with regular light conditions.

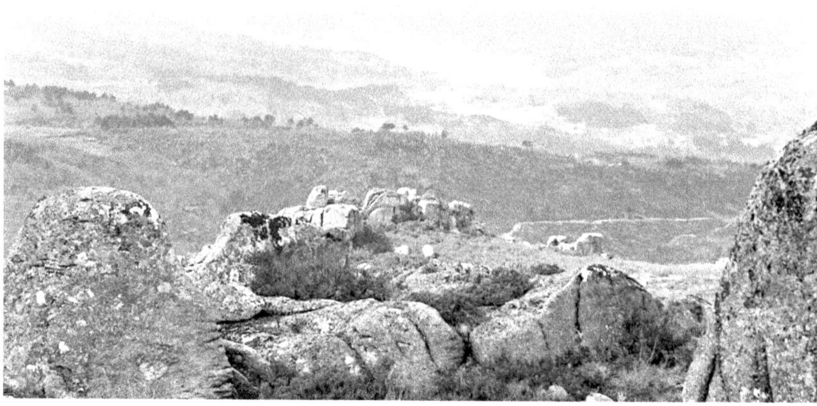

Photograph 3: Serra do Xurés, by Pamela Pérez, 2009 ©

Photograph 4 is an example of one of the new elements introduced: vegetation as a main theme; other themes such as water, rivers and fields are also taken into consideration. The visitors offer a wider representation of the park by introducing all these new views. It is possible that they are not showing the most commonly characteristic parts of the park, but they provide a more complex image of the environment by introducing complementary elements, not necessarily specific to the park, which enlarge our visual experience and power to evoke the landscape.

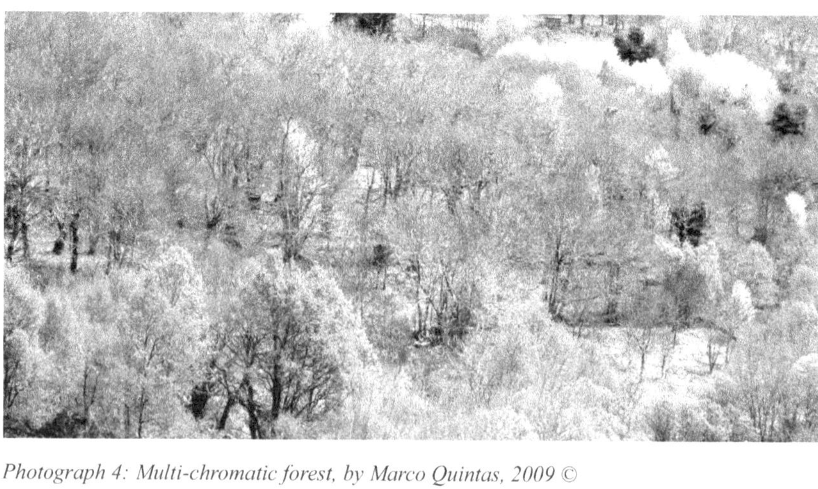

Photograph 4: Multi-chromatic forest, by Marco Quintas, 2009 ©

- **Unpleasant (non assertive icons):** The reason for including this atypical category is to detect the items that should be, in short, improved in the park. As said before, photographs showing disapproval are seldom shared, if not used only specifically for critical purposes. This category gains its entire meaning in connection with what is typical – can something disgusting be typical? The unpleasant has to do with destroyed nature (forest fires), trash and junk in fields (see Photograph 5), buildings in bad condition or the use of something other than vernacular materials, such as granite, tiles and wood, for construction. It is very easy to find these images all around the countryside in Galicia (much less so in Portugal). The ideal image of the countryside based on continuing harmony between humans and the environment here is visually broken by very pragmatic use of material: recycled and cheap construction materials are often used, denying the aesthetic values behind the perception of rural landscape. For local inhabitants this image reproduces nature in production, and the beautiful part is that which produces some tangible good. All the unproductive things, in these terms, have no value. At the same time, touristic representations of a landscape are definitely based on the romantic perception of a rural landscape.

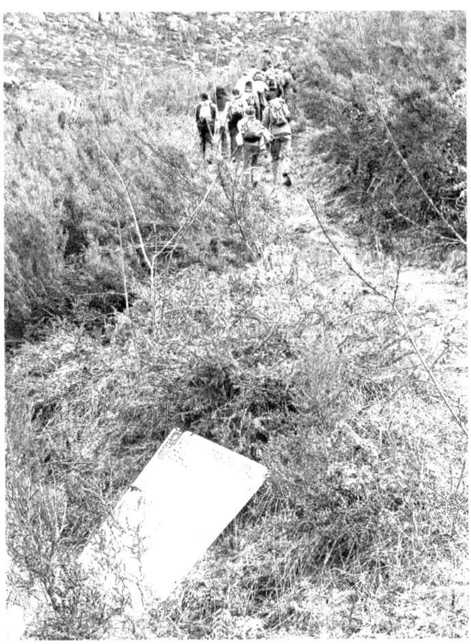

Photograph 5: Junk beside the path, by Óscar Álvarez, 2009 ©

The use, and concept, of environment without an ideal aesthetic value has a great presence in Galicia. Over recent years, a neologism has been created to give a name to this phenomenon: *feísmo* (uglification) (Baamonde). Due its impact in tourism, and the rejection of this practice by part of the population, uglification is a topic of research, internet forums, publications and diverse initiatives (El feísmo en Galicia; Consello da Cultura Galega). For example, there is an initiative started in 2005 and being led by the most important regional newspaper, La Voz de Galicia, which has a section in its online publication for readers to send in photographs showing examples of uglification and vote for the worst.

- **Typical (reinforcing icons):** Typical is not the same as picturesque, because uglification plays an important role in the construction of Galician identity (Gondar). Typical can be either nice or unpleasant, as a blend of positive and negative perceived elements. In this category some images may appear in both of these mainstreams. Animals become present again not as evidence of the biodiversity and biological interest of the park: common cattle, goats and donkeys catch the attention of hikers. These animals are easily found by photographers and are normally contextualised in a surrounding environment with other elements that compose a wider view of landscape. They are common animals in common situations, an image of their presence in everyday rural life; it seems that even more than the animals themselves, the interesting thing is their evocation of farming and rural ways of living. Furthermore, architectonic elements are also taken into consideration: examples of vernacular architecture (houses, granaries, fountains, stone walls, etc.) in use or abandoned (Photograph 6); vernacular buildings with additions of new or recycled materials (uglification); photographs of fields as samples of humanised environments in harmony with nature (in these photographs there are no traces of uglification); close views of vegetation, in many cases of wild yellow gorse – *ulex europaeus* –, which are used in traditional farming and are very common in the mountains and highly appreciated as a wild flower, especially in the Spring when the countryside is covered with its yellow flowers; stone walls, widely used to divide small lots for agriculture; junk (pieces, and refuse, of clothes, engines, construction materials, empty cans, etc.); and wind farms. In recent years wind farms have become a symbol (a quasi-brand) of these mountains and in many areas of the highlands (a large part of the country) the tops of the hills are crowned with windmills for miles. Thus, the wind parks can be seen from great distances, dominating the landscape. In short, with the exception of cattle, all the other themes could also be found in the other two categories ('nice' and 'unpleasant').

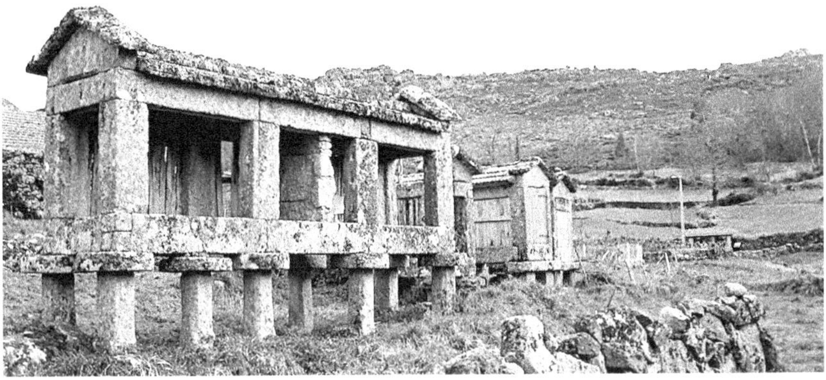

Photograph 6: Granary, by Xonca-Juan Carlos del Carmen García, 2009 ©

Here, the promotional landscape is enhanced with complementary visions, values, meanings and photographs of the environment, and other or richer landscapes are created. Visitors offer their personal interpretations of the canonical landscape, giving an evaluation of its aesthetic value and showing a more complex and useful view of the park, which moves from landscape to multi-landscape.

4. Conclusions

Landscape is more than a social convention related to a given group identity. In tourism, landscape is one of the elements that defines a destination, or in other words, the image of a consumer product. It is a product that any consumer, or even user, may perceive, desire, see, use and experience in an individual way. Visitors create, and often give new senses or introduce foreign elements to conventional meanings. Landscape perception is an individual perception which turns a landscape into a multi-landscape.

Web 2.0 makes it possible to easily identify different perceptions of territory or constructions of landscape by analysing the pictures taken by visitors (not tourism promotors). Such interpretation may present the whole range of territory interpretations to us, and may also contribute to the analysis and evaluation of the aesthetic value of our territory for touristic proposes. Multi-landscapes lead, first, to a redefinition of cultural and natural heritage conception, based on people's perception of value; and, secondly, to the creation of a tool for detecting whatever may be re-considered heritage and what elements of the actual perception of landscape and heritage may be at risk of extinction or being overlooked.

The impact of the sharing possibilities of Web 2.0 in representations of the environment is leading to an enlargement of senses and meanings of landscape. Visitors have introduced new themes, and new views were discovered, giving landscape a dynamic complexity over and above the conventional aesthetic patterns. To the pictorial layer, others layers like the political, private, critical, casual, etc., are added – all of them excerpts of the lived-space. A new assertiveness is being generated thanks to Web 2.0. One that is not separate from the conventional iconic representations of environment, but which better reproduces the complexity of reality and shows the nebulosity of possible knowledge, experience and representation in which underlying values that support landscape representations, bloom. The tourism industry may profit from the emergence of the multi-landscape, and by making environment deeper in meaning we could find a way of preserving natural and cultural heritages.

Acknowledgements

I want to sincerely thank all the people who took part in this project (FotoAcción Xurés) for their collaboration and permission to use their visual material, as well as the people, and my coworkers, from the University of Vigo who were responsible for organising this outdoor activity. I'd like to thank them for their generous help and support.

References

Baamonde, Antón et al.. *Feísmo? Destruír un país: a fin do territorio humanizado. Un novo intracolonialismo*. Ourense: Difusora de Letras, Artes e Ideas, 2006. Print.

Bauman, Zygmunt. *Identity: Conversations with Benedetto Vecchi*. Cambridge: Polity Press, 2004. Print.

Berleant, Arnold. "Towards a Phenomenological Aesthetics of Environment." *Descriptions*. Ed. Don Ihde and Hugh J. Silverman. Albany, NY: State University of New York Press, 2007. 112-128. Print.

Cloke, Paul, and Owain Jones. "Dwelling, Place and Landscape: An Orchard in Somerset." *Environment and Planning A* 33.4 (2001): 649-666. Print.

ConsellodaCulturaGalega."Feísmo."http://www.culturagalega.org/buscas.php?termo=feismo. Consello da Cultura Galega, n.d. Web. 7 Jan. 2010.

De Klerk, Henning. "The Mutual Embodiment of Landscape and Livelihoods: An Environmental History of Nqabara." M.Sc. thesis. Rhodes University, 2007. Print.

El feísmo en Galicia. [...] http://www.foroswebgratis.com/foro-Tema.php?id_foro=48644. N.p., 8 May 2006. Web. 13 Jan. 2010.

Foros do Instituto de Estudos das Identidades. *Territorio, paisaxe e identidade*. Santiago de Compostela: Museo do Pobo Galego, 2007. Print.

Gondar Portasany, Marcial. *Crítica da razón galega: entre o nós-mesmos e o nós-outros*. 2nd ed. Vigo: Edicións A Nosa Terra. Promocións Culturais Galegas, 1995. Print.

Green, Nicholas. "Looking at the Landscape: Class Formation and the Visual." *The Anthropology of Landscape*. Ed. Eric Hirsch and Michael O'Hanlon. Oxford: Clarendon Press, 1995. 31-42. Print.

La Voz de Galicia. "Fotos-denuncia de los internautas." http://www.lavozdegalicia.es/hemeroteca/2005/11/17/100000080950.shtml. La Voz de Galicia, 17 Nov. 2005. Web. 7 Jan. 2010.

La Voz de Galicia. "Nuevos ejemplos de 'feísmo' gallego." http://www.lavozdegalicia.es/fotos/2007/08/23/01109247367CD4A6BC83014618298E82722E.htm. La Voz de Galicia, 10 Aug. 2011. Web. 2 May 2012.

Le Breton, David. *Anthropologie du corps et modernité*. Paris: Presses Universitaires de France, 1990. Print.

Lister, Martin, ed. *The Photographic Image in Digital Culture*. London: Routledge, 1995. Print.

Miyagi, Shunsaku. "Foreword." *Landscape*. Tokyo: Meisei, 1996. 4-5. Print.

O'Reilly, Tim. "What Is Web 2.0. Design Patterns and Business Models for the Next Generation of Software." http://oreilly.com/web2/archive/what-is-web-20.html. O'Reilly. 30 Sept. 2005. Web. 25 May 2005.

Oxford Dictionaries. "Landscape." http://oxforddictionaries.com/definition/landscape?region=us&q=Landscape. *Oxford Dictionaries*, April 2010. Web. 2 May 2012.

Petrarca, Francesco, and James Harvey Robinson. *Petrarch: The First Modern Scholar and Man of Letters. A Selection from His Correspondence with Boccaccio and Other Friends, Dedicated to Illustrate the Beginnings of the Renaissance*. Ed. and trans. James Harvey Robinson. New York; London: G. P. Putnam's Sons, 1898. Print.

Roger, Alain. *Court traité du paysage*. Paris: Gallimard, 1997. Print.

Rose, Mitch. "Landscape and Labyrinths." *Geoforum* 33.4 (2002): 455-467. Print.

Simmel, Georg. *El individuo y la libertad. Ensayos de crítica de la cultura. Las grandes urbes y la vida del espíritu*. 2nd ed. Trans. Salvador Mas. Barcelona: Península, 2001. Print.

Turri, Eugenio. *Il paesaggio degli uomini. La natura, la cultura, la storia*. Bologna: Zanichelli, 2003. Print.

Varela, Francisco J., Evan Thompson, and Eleanor Rosch. *The Embodied Mind. Cognitive Science and Human Experience*. Cambridge, MA: MIT Press, 1991. Print.

David Casado-Neira, PhD, Social and Cultural Anthropology (2001), Professor at the University of Vigo. After working in medical anthropology and embodiment (blood and organ donation), his recent research work is targeted on outdoor tourism, the nature as social construction and the implementation of ITCs for learning outdoors.

Accessible Digital Culture for Disabled People

Marcus Weisen

Director
Jodi Mattes Trust for Accessible Digital Culture
Jouanvins
07160 Le Cheylard
France
marcus.weisen1@gmail.com
www.jodiawards.org.uk

1. Introduction: The digital revolution – freedoms that exclude

The impact of the digital revolution is of a magnitude at least that of the Gutenberg printing revolution. It profoundly affects the way we learn, shape our thoughts and query reality. The production and presentation of culture is being re-shaped in myriad ways. Information is instantly accessible. Almost boundless freedoms have emerged for people to engage with culture in ever new personalised ways through the digital media. But are these freedoms granted in equal measure to all citizen?

In spite of the extent and depth of the changes, there has barely been any discussion in the cultural sector worldwide about the accessibility of:

- digital media for disabled people, as users and creators;
- digital content, such as website information, online collections and learning resources, mobile interpretative tours and cultural productions.

It is widely known that digital technology has enhanced opportunities for learning and employment for a number of disabled people, who enjoy access, to varying degrees, to assistive technology. That's only one side of the picture (and a one-sided one). Far less known is that fact that the web and digital technology are spaces full of pervasive barriers. An audit of web accessibility published in 2005 by the Museums, Libraries and Archives Council, England showed that disabled people face 216 potential stumbling blocks on the average cultural sector web page (Bell et al. 19).

The digital train has left the station with unstoppable power. Fortunately there are stops and stations at which to refuel. It is of paramount importance for the equality of chances that the digital divide between disabled and non-disabled people is being narrowed.

This article examines:

- the accessibility of digital heritage services for disabled people: museum and heritage websites, online cultural collections, and learning resources and on-site digital technology;
- the barriers, the good-practice, as well as institutional and creative ways forward;
- international policies for the human, cultural and digital rights of disabled people, which cultural ministries worldwide like not to think about;
- the cultural exclusion of disabled people, which is compounded by digital exclusion in the cultural sector.

2. The cultural exclusion of disabled people

2.1 A history of cultural apartheid

It is still uncommon to talk about the 'cultural equality of disabled people'. Not so long ago talk about making museums and cultural services accessible to disabled people has been seen an optional and benevolent act of charity (Delin 6).

Lack of access was seen as the disabled person's problem. The idea, that cultural venues, as a service to the public, have a responsibility to welcome all, is far from being universally embraced in the cultural sector.

Newly built museums and museum extensions provide a compelling example of cultural apartheid being systematically practised against disabled people. Billions of Euros have been spent over the past decade on hundreds of new museums and their exhibitions. Only a small minority have planned even a small measure of intellectual access for casual visitors who are visually impaired, deaf or who have learning disabilities (Weisen 18). Those which plan some degree of shared experience with disabled people, often do not care to evaluate what quality of service is provided. In terms of institutional knowledge management, this is a culture of neglect.

2.2 Barriers to cultural experience

2.2.1 A whole organisation approach to accessibility

The accessibility of cultural venues is a complex and multi-dimensional reality. For accessibility to work optimally, consideration for accessibility needs to be seamlessly integrated into all aspects of the museum experience: visitor information – including via digital media; the physical environment, signage, exhibitions, interpretation and, last but not least, institutional attitude and staff welcome. Of course, direct knowledge of disabled people is key to good service provision. Budgeting and resource allocation which is inclusive of disabled people is fundamental. To be effective, accessibility needs to be planned from the outset into every project. In short, these are some of the key elements of a holistic approach to accessibility. Digital media need to be part of it and allowed to deploy their full emancipating potential. The failure to embrace such an approach will result in countless unnecessary access barriers being erected.

2.2.2 Experiences of disabled people

Let's listen to the experiences of a few disabled friends and colleagues of mine:

> I will never go back to a museum. There is nothing about my culture. It is as if deaf people did not exist.
>
> The theatre is new and physically accessible, but staff behaves as if I came from another planet. I prefer to go to the other theatre, even though staff needs to carry me up a few stairs.
>
> The person at reception was hopeless! They didn't have any idea about what services they had available for visually impaired people.

Getting to know the experiential ways and requirements of disabled people is key to dismantling barriers. For example, cultural organisations remain largely unaware of the requirements of hard of hearing people. They are by far the largest group of disabled people and, in many countries, hearing loops have become very affordable (150-200 euros). Yet, less than 1/3 of reception desks had a hearing loop system (Scope).

Barriers are inter-dependent and repeated experience of unnecessary barriers leads to frustration, anger, resignation and finally to cultural exclusion.

2.2.3 Museums worldwide do not demonstrate a systematic approach to access

The few examples above – taken from many possible ones – bring home awareness that the creation of barrier-free spaces and cultural experiences requires a systematic approach. The uncomfortable truth is that such an approach is lacking worldwide. Less than half museums, libraries, archives in the UK, for example, are building commitment to accessibility into their core budgets, and this situation is typical for global practice (Bell, Matty) – and these figures are optimistically skewed, as they derive from a small, responsive sample.

2.3 The human and cultural rights of disabled people: The policies states and cultural organisations still like to forget

2.3.1 Cultural participation is a human right of disabled people

The idea that disabled people have cultural rights is as yet to become an integral part of the philosophy, vision and practice of most cultural organisations worldwide (Weisen 245).

The cultural rights of disabled people are enshrined in the Universal Declaration of Human Rights (article 27.1):

> Everyone has the right freely to enjoy the arts and the cultural life of the community…

The UN Convention on the Rights of People with Disabilities, which came into force in 2008, explicitly recognises the cultural rights of disabled people for the first time: "States Parties recognize the right of persons with disabilities to take part on an equal basis with others in cultural life, and shall take all appropriate measures to ensure that persons with disabilities…" (article 30). The Convention is also explicit for the first time in recognising deaf people as a linguistic minority (as countries such as Finland and the UK have done earlier). This is important, because it builds the case for cultural organisations providing services in Sign Language, live and via digital media.

2.3.2 The good European policies and the lack of commitment to implementing them

The European Union and the Council of Europe have developed some of the most advanced cultural rights policies for disabled people, but the signs are that these are not being implemented. In 2003, the Council of the European Union passed the 'Council Resolution of 6 May 2003 on accessibility of cultural infrastructure and cultural activities for people with disabilities'and agreed to monitor measures taken by member states by the end of 2005. To date no such progress monitoring exercise has been undertaken.

The Council of Europe Disability Action Plan (2006-2015) on full participation of people with disabilities in society emphasises equal opportunities in culture (chapter 3.2, p. 13). It calls on member states to "start with an evaluation of their existing disability policy programmes and identify in which areas progress has yet to be made […] and which specific actions will have to be carried out" (chapter 1.5, p. 8). To date, I know of no single Council of Europe member state, that has undertaken an evaluation of its cultural equality policy programmes for disabled people!

The Council of Europe Recommendation No. R (92) 6 on the independent living of disabled people (1992) is the first international policy to call for the implementation of the cultural rights of deaf and disabled people:

> Government institutions, leisure and cultural organisations should develop comprehensive access policies and action programmes designed to significant and lasting improvements for all people with disabilities (chapter VIII, section 8.5).

This Recommendation is conceptually strong, because it recognises the necessity for systemic and systematic change. As such it is unsurpassed in its uncompromising clarity. No national government can claim to have implemented this recommendation, as none has even developed a set of tools with which to measure whether 'significant and lasting improvements' have taken place.

E-accessibility, which covers all areas of life and is of profound significance, is one area in which the EU has been continuously active. It has been policy since 2002 that "public sector web sites [...] must be designed to be accessible." (European i2010 initiative on e-Inclusion, 4). Information on national policies regarding web accessibility can be found on the Web Accessibility Initiative of the World Wide Web Consortium.

3. Cultural websites: The experience of disabled people

The Museums, Libraries and Archives Council, England, is one of the few strategic cultural bodies to have commissioned an audit of sector web accessibility (Petrie, King, Hamilton). Although published in 2005, it remains unusually interesting because of its strong user focus. In addition to customary automated testing, testing with disabled users took place in realistic settings. 300 museums, libraries and archives websites in England and 25 large museums overseas were audited. As web accessibility has not significantly improved since, the findings remain relevant today.

Only 3% of websites met world wide Web Content Accessibility Guidelines (WCAG1) Level AA, the EU e-government requirement websites. The average number of instances of WCAG1 checkpoint violations that occurred per page was 56.9.

Table 1: Average number of violations and frequency per page.

Type of checkpoint error	Violations	Instances
Automated	5.9	56.9
Manual 'warning'	34.3	159.0
Total	40.2	215.9

The average museum, library and archive home page may have as many as 215.9 instances of potential stumbling blocks to users. This is a particularly worrying situation when we consider that many of the problems users actually encounter when using web sites are 'warning' violations of the checkpoints that do indeed require manual checking.

15 disabled User Panel members attempted a total of 120 tasks with 20 web sites selected for in-depth testing (20 web sites x 2 tasks per web site x 3 evaluators per web site = 120). Finding out opening times and access information were typical tasks.

The Panel members succeeded in 75.6% of the attempted tasks and failed in 24.4% of them. Failure to complete tasks was not attributed to a minority of the participants, but from a broad cross-section of each User Panel.

Table 2: Disability group task success.

User Group	Tasks succeeded
Blind	66.7%
Dyslexic	82.5%
Partially sighted	77.5%

The Panel members, when asked about the extent to which their impairments were taken into account, gave a mean rating of 3.4 on a scale of 1 to 7.

The problems observed by the research team and the Panel members were collated and categorised. Overall, 189 instances of problems were identified. 147 (78%) directly related to WCAG1 guidelines, whilst 42 (22%) were not covered by these. Below are the most common problems users encountered. These problems undoubtedly explain the failure rates summarised earlier.

Table 3: Most frequent user testing problems.

Key problems experienced by User Panel as a whole	**Instances**	**In WCAG1**
Target of links not clearly identified	30	Yes
Information presented in dense blocks with no clear headings to identify informational content	17	Yes
Inappropriate use of colours and poor contrast between content and background	14	Yes
Navigation mechanisms used in an inconsistent manner	13	Yes
Links not logically grouped, no facility to skip navigation	10	Yes
Text and images do not increase in scale when browser option selected	7	Yes
External information and navigation on page, not associated with page content	6	No
Important information not located at top of list, page, etc.	6	Yes
ALT tags on images non-existent or unhelpful	6	Yes
Graphics and text size too small	5	No
Distraction and annoyance caused by spawned and pop-up windows	5	Yes
Labels not associated with their controls	5	Yes
Images and graphical text used in-place of plain text	5	Yes

These 13 problems constitute 68% of the total number of problems uncovered during the user testing evaluations. Over half of these problems relate to matters regarding orientation and navigation.

Poor page design (in terms of layout) led to a recurrent orientation problem for all the user groups involved in the evaluations. Both the research team and the members of the User Panel considered many sites to have overly complex and "cluttered" pages with dense blocks of text. No clear indication of main headings, secondary headings, etc., was a recurring problem. While sighted users could infer some of this logic from text sizes, colour coding, etc., blind users did not have access to this and so pages were deemed "illogical", meaning they lacked a logical structure.

Ambiguously named links that led to unexpected content were responsible for many of the navigation problems users encountered, e.g. opening times were often found under 'Contact Us'. As one dyslexic user commented "...important information like opening times and disabled access should not be hidden under other obscure titles."

Colour scheme and contrast used for page designs accounted for many of the complaints from the dyslexic and partially sighted users. The colour scheme often affected these users' abilities to perform tasks, particularly when the contrast between the text and the background was inadequate. Pale text on pale backgrounds was a common problem. Moreover, different users benefit from different colour schemes. Although colour schemes can be changed by users (e.g. by attaching their own style sheets to their browser) very few users seemed to be aware of this.

No *'skip navigation'* link at the top of pages enabling blind users to jump to the main content of a page (by-passing the page's top navigation) was a specific problem for Panel members who used screen readers. When such links were missing, blind participants were compelled to listen to the navigation elements that commonly appeared at the top of pages: repetitive information they often describe as audio "clutter" and find very frustrating and exhausting.

External navigational links. Numerous academic and local authority museum, archive and library sites are integrated (relatively) into a host institutions external site. This caused confusion to all user groups. The user was commonly unaware that the external navigational links did not directly relate to main content of the page: "keeps giving me information about other things [...] information about Civic Centre. Think I must keep wandering off." (Comment by partially sighted participant)

Perhaps unsurprisingly, many of the positive aspects reported by Panel members were the opposite of the problems outlined above. For example, partially sighted participants appreciated "good use of colours to highlight visited links". Blind users enjoyed logically structured pages, and as one user put it: "Proper links, labelled individually and properly mean no trawling is necessary." The other user groups also liked sites that had clear navigation mechanisms, logical page layouts, clear contrast, reasonably sized text and straight-forward language.

The MLA report is a prime example which shows how important user involvement in testing is. It identifies significant room for improvement. It also affirms (against a simplistic understanding of web accessibility) that an accessible web site does not need to be plain, boring or text only, just creatively designed for all. The needs of disabled users must be addressed with appropriate alternatives for users who due their disability and/or assistive technology cannot efficiently interact with specific features. It advocates the use of exciting technologies and designs in the construction of web sites.

4. The accessibility of cultural content: A unique role for cultural websites and digital media

4.1 Promoting accessible digital media, as well as intellectual access to collections

The MLA audit highlighted that museums and heritage websites can make a "unique contribution" through accessible presentation and interpretation of collections and learning resources, for example by providing descriptions of digital images for visually impaired people or information in British Sign Language (MLA 34).

However, less than a handful of sites provided these services, and when so, these were of very modest scope. No site audited demonstrated engagement with users with a learning difficulty – who are likely to find scholarly and curatorial language a barrier. The number of museums and heritage websites which provide any of these services remains shockingly small in 2010.

The Jodi Mattes Trust works to change this. It gives the annual Jodi Awards for accessible digital media to museums, libraries, archives and heritage sites. It celebrates best practice and promotes the human, cultural and digital rights of disabled people. Given since 2003 in the UK, the trust launched an International Award for best website in 2009. The Jodi Awards website provides a range of informative best practice case-studies of Jodi Awards 2009 winners (Jodi Mattes Trust). Four case-studies of previous winners and a case-study of the Cité des sciences et de l'industrie, which all have memorable qualities, are presented below.

4.2 Best web site practice: Tate i-Map – creative, attractive, pedagogical[1]

Tate i-Map, Jodi Award 2006 winner, still sets a standard for global best practice (Tate Modern). This site does what seems impossible to many people by making modern art – and its key concepts, accessible to blind and partially sighted people.

1 www.tate.org.uk/imap/

Created to provide learning resources for the Picasso-Matisse exhibition (and expanded since), it is one of first to describe collections for visually impaired people. Caro Howell, then Manager Special Projects, and currently Head of Education at the Whitechapel Art Gallery in London, describes how the project originated from a sense of frustration at the lack of learning materials for visually impaired pupils. She was horrified when during a school visit a visually impaired pupil was asked by her tutor to draw a dragon whilst the rest of the class was to be initiated to the fine concepts of the emergence of abstract art.

Tate i-Map represents paintings in high colour contrast, making them easier to see for many partially-sighted people. Colour contrast is being used as an integral part of the presentation and analysis of the works of art, making compositional elements stand out.

The clever use of animation ingeniously breaks the 'no flash' dogma propagated during the early years of the WCAG1. Moving from de-constructed fragments to the re-constructed whole of a painting, the slow motion animations little by little builds up a compelling picture of the transformation of figurative into modern art. This enriches every viewer's understanding. Visually impaired viewers can replay the animation when this helps get the picture, just as they might replay favourite scenes from a DVD at home.

The approach taken to provide access for totally blind viewers is equally innovative. The black on white outline drawings of the paintings can be printed out in schools and centres for visually impaired people, as well as blind people who have access to a 'tactile photocopier' and 'swell paper'. In the first month of the website some 3,000 drawings were downloaded, demonstrating that a demand exists. Tactile drawings are not new in themselves – they have been used for some hundred and fifty years in education. What is new is their availability online in the cultural sector. On their own they make no sense, but Tate i-Map also provides essential descriptive and interpretative information.

4.3 Best digital media practice: Bantock House, a holistic approach to engaging with deaf audiences[2]

Wolverhampton Arts and Museum Service, Jodi Award 2007 winner, remains a prime example of a museum taking a systematic approach to removing accessibility barriers for deaf people. Accessibility is usually patchy and the service offered highly restricted. Opportunities for casual visits, just 'dropping in', remain all too rare, yet they really are a condition for enjoyment of cultural freedoms. Poverty of choice is a deterrent for may disabled people.

---

2 www.wolverhamptonart.org.uk/bantock/the_house/bsl

Wolverhampton Arts and Museum Service exemplifies how a medium-size museum can deploy limited resources in a thoughtful way to provide choice and a 'portfolio' of services. The service embarked on a project to produce a PDA guide with British Sign Language (BSL) of Bantock House, which would enable deaf visitors to visit at a time of their choosing. In the process, the service quickly became aware that the PDA guide would need to be publicised in Sign Language! Website information in BSL was produced. The videos for the PDA and the website have been filmed by a local company of deaf people who are all BSL users, communicating effortlessly with member of the deaf community.

Aware that the PDA guide, which offers the freedom to 'drop in', only offers one story, regular guided BSL tours have been introduced to widen access to the collections in personalised ways. In 2008, Wolverhampton Arts and Museums Service organised an exhibition of deaf history about the lives of deaf people in Wolverhampton. In the process, it became quite possibly the first museum service in the UK ever to create archival footage in BSL – raising important questions of collections policies and representation of deaf and disabled people in collections. It provided a space in which local communities could get to meet one of least known and most segregated of its communities. A deaf person was employed part-time to market Wolverhampton Arts and Museum Service's offerings to deaf community people – someone who knows how to communicate with the target audience.

Wolverhampton combines a whole range of approaches, and involved deaf people directly in service development. 'Traditional' museum offerings such as live guided tours and a social history exhibition go hand in hand with the use of digital media – a PDA, website information and archival social history footage.

4.4 Prisoner 4099, The National Archives, Jodi Award winner 2008[3]

This easy to use site breaks new ground. Young visually impaired people created learning content for both informal and curriculum learning to be enjoyed by everyone. The National Archives made the archival documents on which the visually impaired pupils based their radio play accessible by transcribing them.

At every stage of the website design staff at the museum were insistent that the website must be accessible to disabled people. They were forthright in challenging the web designers about their knowledge.

Let's reflect: A truly inclusive effect was achieved, by reversing roles. Usually on the receiving end, people with disabilities were put in the driving seat.

Even in big, hugely busy institutions personal commitment matters and can bring about an altogether new quality of service.

---

3 www.nationalarchives.gov.uk/education/prisoner4099/

4.5 THE British Museum deaf school project, Jodi Award winner 2009[4]

In this project young deaf people produced signed curriculum resources for young deaf people, working with Frank Barnes School and deaf led media company Remark. The collaboration brought together a creative and appropriate mix of users, artists and designers with expertise in the area of BSL. The pupils confidently present the collections.

It is well worth reflecting on the importance of Sign Language. Learning resources and visitor information in Sign Language remain extremely rare in museums, as are curriculum resources in Sign Language. We can speak of a situation of cultural segregation.

The new UN Convention on the Rights of Disabled People recognises deaf people as a linguistic minority for the first time. This encourages use of sign language, at the least in medium size and large projects. The exclusion of Sign Language from public spaces is, in the first instance, grounded in the failure to recognise deaf people as a linguistic and cultural minority – something scientist Oliver Sacks came round to understand 20 years ago and described in his book *Seeing Voices*.

An argument is made by some that deaf people read subtitles. Yet, this is only partly true and many born deaf people have very poor alphabetic literacy skills. A number of young deaf people prefer Sign Language by far. Irrespective of what the law and policy might say, we should welcome deaf people on their own terms wherever possible. It all starts with inclusive budget setting. Personalisation and access requirements invite us to plan communications and interpretation strategies which provide both Sign Language and closed captioning.

4.6 Cité des sciences et de l'industrie – *a leader*

When the *Cité des sciences* opened in 1986 in Paris, it had a Disability Access Unit with four full-time members of staff, which has now grown to eight (that is more than the total number of Disability Access Managers of all other national museums in Paris, London or New York). Among them are two deaf and one blind member of staff. Several others are conversant with French Sign Language (LSF).

All exhibitions, temporary and permanent, have to provide a degree of accessibility to visually impaired people. Curators and designers have to involve the Access Unit at planning stages – and though consultation is not always without tensions, a sophisticated culture of accessible exhibition design has grown which remains without a par.

4 www.britishmuseum.org/learning/schools

In 2010, the temporary exhibition 'Earth' showed a DVD about the journey of stones from mountains to the building site. It is mainly a silent movie. The visuals are self-explanatory and have cartoon-like clarity of communication. They put deaf and non-deaf visitors pretty much on an equal footing. Six screens with silent film and sub-titles have been audio described for visually impaired viewers.

A number of tactile models explain the physical properties of traditional building materials all over the world with Braille captions. Here too a creative exhibition designer was at work to create elegant design solutions. They designed tactile models to communicate scientific knowledge to all.

On the next floor, in the temporary exhibition 'Bon appétit', deaf actors perform in a hilarious movie raising questions about the healthy food and junk food. La Cité not only provides Sign Language on screen more abundantly than any other museum in the world, it creates DVD scripts, which give visibility to deaf actors. At last year's temporary children's exhibition about sexuality, deaf actors explained human reproduction on screen. Non-deaf people were provided with headsets. It became the most popular exhibit of the show. A blind twelve-year-old child said: "This is great, for once I could enjoy the exhibit without my mum having to describe it for me."

La Cité des sciences also produces tactile books with audio sound tracts on DVD.

Reflecting on this experience, we can see that twenty five years of commitment and experimentation have put la Cité des sciences at the forefront of accessible exhibition design. Its exhibitions show a new emerging quality of accessible and inclusive design that reveals just how creative engagement for the cultural rights of disabled people can be. If more join in, the inclusive design harvest will eventually turn abundant. Designers and new media professionals, take note!

5. Conclusion: Accessible digital culture – human right, political need, ethical duty

This article showed that:

- the participation of disabled people in cultural life is a recognised human and cultural right;
- in spite of progress, disabled people continue to face significant barriers to culture, and specifically so in digital media;
- national ministries for culture do not implement the policy commitments and aspirations for cultural equality they have expressed in policy documents;
- few cultural organisations make disability equality and access integral to the whole planning process – and therefore are to be held responsible for the creation of vast numbers of unnecessary access barriers;

- for these reasons, there are instances in which disabled people are clearly being denied their cultural rights.

This article also:
- highlighted the largely untapped potential of digital media to be deployed in the service of the cultural equality of disabled people;
- drew attention to inspiring best practice;
- affirms the belief that access and elegance, access and creativity are not opposites, but go hand in hand;
- holds up the unshakeable belief that culture without concern for sharing its cultural wealth is uncivilised;
- affirms the urgency for action.

It concludes with a few general recommendations, expressed in a spirit of burning patience and burning impatience:
- ministries for culture should aim high and develop strategic implementation programmes for the cultural and digital rights of disabled people, which bring 'lasting and significant change' as called for by Council of Europe Resolution R(92)6;
- disability equality should be a criterion for all cultural funding;
- cultural organisations should make disability equality and access integral to all their work, including the budget;
- equality and access for disabled people should be in the script of every brief for digital projects;
- disabled people should be involved in the process;
- the cultural sector should learn from the further education experience and further adapt the holistic model of web accessibility developed by Kelly, Phipps and Howell.

Whatever our role and contribution, we can take inspiration from Ghandi's humanising spirit of service. It is within our power and inner freedom to apply some of our passive admiration for Ghandi into the active every-day life of our cultural institutions:

> A customer is the most important visitor to our premises. He is not dependent on us. We are dependent on him. He is not an interruption on our work. He is the purpose of it. He is not an outsider on our business. He is part of it. We are not doing him a favour by serving him. He is doing us a favour by giving us opportunity to do so.

References

Bell, Janet A., Simon Matty, and Marcus Weisen. "MLA Disability Survey 2005." London: Museums, Libraries and Archives Council, 2005. Web. 28 Feb. 2010. www.nemo.org/fileadmin/Dateien/public/topics/Disability_and_museums/disability_survey_2005_summary_10447.pdf.

Commission of the European Communities. "eEurope 2002, Accessibility of Public Sector Websites and their Content." 2001. Web. 28 Feb. 2010. http://eur-lex.europa.eu/LexUriServ/LexUriServ.do?uri=COM:2001:0529:FIN:EN:PDF.

Commission of the European Communities. "European i2010 initiative on eInclusion. "Web. 28 Feb. 2010. http://ec.europa.eu/information_society/activities/einclusion/policy/i2010_initiative/index_en.htm.

Council of Europe. "Disability Action Plan 2006-2015." 2006. Web. 28 Feb. 2010. www.coe.int/t/e/social_cohesion/soc%2Dsp/Rec_2006_5%20Disability%20Action%20Plan.pdf.

Council of Europe. "Recommendation No. R (92) 6 of the Committee of Ministers to Member States on a Coherent Policy for People with Disabilities." 1992. Web. 28 Feb. 2010. www.handicapincifre.it/allegati/RECOMMENDATION_R(92)6.htm.

Delin, Annie. "Disability in Context". Disability Portfolio. London: Museums, Libraries and Archives Council, 2003. Web. 28 Feb. 2010. www.mla.gov.uk/what/support/toolkits/libraries_disability/find_out_about_disability.

European Council. "Resolution of 6 May 2003 on Accessibility of Cultural Infrastructure and Cultural Activities for People with Disabilities." Council of the European Union, 2003. Web. www.rech2006.com/download_files/resolution_en.pdf.

European i2010 initiative on e-Inclusion. Communication from the Commission to the European Parliament, the Council, the European Economic and Social Committee and the Committee of the Regions. Web. 28. Feb 2010. http://ec.europa.eu/information_society/activities/einclusion/policy/i2010_initiative/index_en.htm

Jodi Mattes Trust for Accessible Digital Culture. Web. 28 Feb. 2010. www.jodiawards.org.uk/awards.

Kelly, Brian, Lawrie Phipps, and Caro Howell. "Implementing a Holistic Model of E-Learning Accessibility." Eds. John Cook, and Denise Whitelock. *Exploring the Frontiers of E-Learning: Borders, Outposts and Migration. ALT-C 2005 12th International Conference Research Proceedings. ALT Oxford.* UKOLN. 2005. Web. 28 Feb. 2010. www.ukoln.ac.uk/web-focus/papers/alt-c-2005/.

Petrie, Helen, Neil King, and Fraser Hamilton. *Accessibility of Museum, Library and Archive Websites: The MLA audit.* London: MLA; City University 2005. Web. 28 Feb. 2010. www.egovmonitor.com/reports/rep11278.pdf.

Sacks, Oliver. *Seeing Voices: A Journey Into the World of the Deaf.* New York: Harper Perennial. 1990. Print.

Scope – Working for People with Cerebral Palsy. Survey of Accessibility to Services. London, 2000. Print.

Tate Modern. Web. 28 Feb. 2010. www.tate.org.ukimap.

United Nations. "Convention on the Rights of Persons with Disabilities." UN, came into force 8 May 2008. Web. 28 Feb. 2010. www.un.org/disabilities/default.asp?navid=13&pid=150.

United Nations. "Universal Declaration of Human Rights." UN, 1948. Web. 28 Feb. 2010. www.un.org/en/documents/udhr/.

Web Accessibility Initiative. "Policies Relating to Web Accessibility".Web. 28 Feb. 2010. www.w3.org/WAI/Policy/.

World Wide Web Consortium. Web. 28 Feb. 2010. www.w3.org/.

Weisen, Marcus. "How Accessible Are Museums Today." Ed. Helen J. Chatterjee. *Touch in Museums: Policy and Practice in Object Handling*. Oxford; New York: Berg, 2008. 243-252. Print.

Wolverhampton Arts and Museum Service. Web. 28 Feb. 2010. www.wolverhamptonart.org.uk/bantock.

Marcus Weisen is a Museum Access Consultant, a Collaborator of the European Centre for Cultural Accessibility (www.culturalaccessibility.org) and Director of the Jodi Mattes Trust for accessible digital culture (www.jodimattestrust.org.uk). He was Health and Disability Adviser, Museums, Libraries and Archives Council (2002-2007) and Arts Officer, Royal National Institute of Blind People (1987-2002).

A Short History of Participation in the Cultural Realm

Nico Carpentier

Vrije Universiteit Brussel
Charles University in Prague
nico.carpentier@vub.ac.be

1. Introduction

Roland Barthes' *Image Music Text* contains the seminal essay *The Death of the Author*, which pointed to the convergence between the producers and receivers of discourses at the level of interpretation. The death of the Author was a metaphor, not be taken literally, implying that there was no privileged vantage point that fixed the interpretation of a text. But it also referred to structural power changes in society, where members of cultural elites could no longer claim control over their writings. "Ordinary" readers became (seen as) more and more capable of producing their own interpretations, which might structurally diverge from the intentions of the author. As we have more recently witnessed an increased convergence between the producers and receivers of discourses at the level of the production process, we could say the Author died a second time. The old Author is no longer solely in control of the production process, as the "produser" (e.g., Bruns) has overcome the rigid separations between both categories. Again, this is seen as a major step towards the democratisation of our cultural realms.

There are a number of problems with this type of argument. First of all, the argument tends towards an individualised interpretation of the social, which leads to a downplaying of societal structures, including the importance of organisational structures in providing cultural elites with safe havens, and the importance of discursive structures like professional identities and audience identities. Obviously, these structures are interdependent, as institutions act as discursive machineries, producing identities, and professional identities are driving forces for the functioning and legitimisation of cultural institutions. These structural components make the Author more resistant than it seems. S/he has indeed found shelter in a series of organisations and institutions, protected by their professional structures and organisational cultures that provide networks of support and resources. The Author is also resistant at the cultural-discursive level, as the contemporary subject positions related to the Author (or the many cultural professionals), turn out to be more rigid than expected (and sometimes desired). In other words, cultural professionals' identities remain embedded within hegemonic discourses on management, autonomy and expertise.

But this is not the only problem with the death-of-the-Author argument. Especially in the 1990s and 2000s we have witnessed a strong revival of this argument, connected to the changes in the communicational landscapes. The popularisation of the Internet, with all its potentials for interaction and participation fed into the cultural democratisation argument, combined with the belief that these changes were new. This claim for novelty is highly problematic, as it tends to ignore the history of cultural participation, which stretches out much further than utopian ICT theories want us to believe. This article wants to combine both critiques, by looking at the 20th century history of participation within the cultural realm. Arguably, the arts played a significant role in producing these cultural-democratic discourses and practices, and for that reason this article will first discuss this history, only to move towards the debates on museums and participation afterwards. This historical overview aims to show that one should be careful with launching claims of novelty when discussing cultural participation. It will also allow me to emphasise the importance of discursive structures as conditions of possibility for the organisation of participation. Over time, the articulation of participation itself, and the discourses on how and to what degree power imbalances should be equalised, have put their mark on participatory theories and practices.

2. Participation and the world of the arts

In the world of the arts, participation has been thematised and practised in many variations,[1] although what is termed 'participatory art' can hardly be considered a canonised art movement (any longer). Obviously, the artist has a strong power position in the creation of the artwork, but as Boris Groys (20) remarks, this power position remains incomplete ever since art was secularised: "No modern artist would expect anyone to kneel before his work in prayer, expect practical assistance from it, or use it to avert danger." For its appreciation, secularised art has become dependent on arts markets (in many different forms and shapes), which generate financial value and on what Groys calls 'public taste', which cannot be equated with financial value. As Marcel Duchamp wrote:

> In the last analysis, the artist may shout from all the rooftops that he is a genius: he will have to wait for the verdict of the spectator in order that his declarations take a social value and that, finally, posterity includes him in the primers of Artist History. (77)

---

1 One example is the exhibition *The Art of Participation. 1950 to Now*, which was organised by Rudolf Frieling for the San Francisco Museum of Modern Art, and which was on view from November 8, 2008 until February 8, 2009. The exhibition catalogue by Frieling, Groys, Atkins and Manovich has been very helpful in providing both reflexive articles and examples.

Nevertheless, artists obviously remain crucial actors in the creative-artistic process, which unavoidably results in a more passive position for the audiences of arts works. These audiences stroll (quietly) through museums and galleries, and (even more quietly) attend performances and screenings. This distance also allows them to pass their judgements from a position which is external to the artwork.

But in a number of cases art has also offered reflections on the always problematic relationship between the artist, the artwork and its audiences. In the case of participatory art, these reflections are translated into revisions of the traditional passive position of the audience, implicating the audience into the artwork. Here, the strong position of the artist, as creator of the artwork, does have an impact on the degree of participation that is allowed for. In many cases, members of the audience are interacting with an already produced work of art, are given guidelines on how to perform to generate or complete the artwork, or act in ways that are then incorporated into the artwork. Secondly, one should also not remain blind for the strengthening of the position of the artists through audience participation, as Groys points out:

> When the viewer is involved in artistic practice, every piece of critique he utters is self-criticism. The decision on the part of the artist to relinquish his exclusive authorship would seem to primarily empower the viewer. This sacrifice ultimately benefits the artist, however, for it frees him from the power that the cold eye of the uninvolved viewer exerts over the resulting artwork. (21)

An important starting point for these reflections on participatory art is Richard Wagner's essay *The Art-work of the Future*, who in his plea for the *Gesamtkunstwerk* (or the total artwork) accuses the arts of egoism, partially because the arts has divided itself into a variety of genres and became disconnected from the people, which are seen as the source of all creativity: "The Art-work is the living presentation of Religion; – but religions spring not from the artist's brain; their only origin is from the Folk [das Volk]." (Wagner 18) In his essay, Wagner calls upon his fellow-artists to return the arts to the people, in producing the artwork of the future:

> But to you I turn, – in the same sense as the Folk, albeit of necessity in your own mode of utterance, – to you, ye prudent men and intellectual, to offer you, with all the People's open-heartedness, the redemption from your egoistic incantations in the limpid spring of Nature, in the loving arm-caresses of the Folk – there where I found it; where it became for me my art-instructor; where, after many a battle between the hope within and the blank despair without, I won a dauntless faith in the assurance of the Future. (11)

Wagner's total artwork attempts to reconfigure the arts' position towards the people, supported by the argument that "The richest procreative force lies therefore in the utmost multiplicity" (ibid. 9). But as Groys (23-24) remarks, Wagner's attempt to establish an artistic fellowship with the people and to undermine the author's power, remains ambivalent, as the author remained in control of the stage and the codes of the Wagnerian operas remained almost impenetrable.

Later, twentieth century art movements like Futurism, Dadaism and Surrealism used provocation and scandal to decrease audience passivity by transforming the audience member into a "hostile participant, provoked, attacked and beaten by author and actors" (Melzer 43). Jackson and Lev-Aladgem (2004) describe how Dada artists used strategies like putting glue on seats, selling the same entrance ticket to several people and pinching (female) visitors. At the end of the 1930s, Antonin Artaud developed the *Theatre of Cruelty*, which used similar strategies to decrease the distance between actors and audience, by confronting them with extreme sounds, light and gestures. But especially in the 1960s, a series of arts movements strongly emphasised the concept of participation. Bishop (15) mentions three movements: Situationism (France), Happenings (United States) and Neo-Concretism (Brazil). In France, the Situationist International, with Guy Debord as one of its main protagonists, emphasised the connection between art and (radical) politics, and critiqued the impact of capitalism on everyday life, leading for instance to pseudocommunication and a lack of participation. In the so-called situations (or "collective environments, ensembles of impressions determining the quality of a moment" (Debord 98)), from which the Situationist International derived its name, more intense life experience would become possible. A situation is "made to be lived by its constructors. The role of the 'public', if not passive at least a walk-on, must ever diminish, while the share of those who cannot be called actors but, in a new meaning of the term, 'livers' [viveurs], will increase." (Ibid.)

Happening, which first developed in the United States, "aimed to manipulate creatively the relationship between the presented materials, performers and spectators … Spectators became 'participants' who by carrying on simple 'tasks' and 'activities' aided in the creation of metaphors." (Jackson and Lev-Aladgem 209) Allan Kaprow, who is credited for having staged the first happening (Zimbardo, "Allan Kaprow" 102), published a series of guidelines for happenings, which emphasised that the line between art and life should be kept fluid, the sources (of the happening) should originate from outside the arts, the happening should take place in widely spaced locales, time should be discontinuous and variable, the happening should be performed only once, and – most importantly in this context – "audiences should be eliminated entirely" (Kaprow 713). Kaprow (ibid.)

explains: "A Happening with only an emphatic response on the part of a seated audience is not a Happening but stage theater," but he also resists the idea of having unprepared "participants" being submitted to the abuse of the artists. In contrast, Kaprow (714) prefers respectful events with participants that are willing and committed; at the same time the participants' professional talent is not a requirement. He furthermore nuances the idea that happenings necessarily have to have active audiences: participants might not always know that they are part of the happening – Kaprow mentions the example of a butcher that sells meat to a customer-performer – and a happening can also be staged for the audience just to watch it.

Quite a number of artists have used the medium of the happening, like for instance those affiliated with Warhol's Factory, or the Fluxus group, but these groups also used other formats to facilitate audience participation. In *Water Yam* and *Fluxkit*, George Brecht for instance produced collaborative toolkits, that included instructions for the actions of participants, but also for the creation of objects (Frieling, "Towards Participation in Arts" 41). Nam June Paik's *Participation TV* allowed the visitor to produce voice-generated television images, what Zimbardo ("Nam June Paik") called "unpredictable explosions of lines," and Yoko One staged the *Cut Piece* performance, where audience members were invited on stage to cut off pieces of her clothes until she was nearly naked. In contrast to the Fluxus group's work, Andy Warhol's participatory art was more aimed to attempting to enlist others "to work towards the mass production of Warhol images" (Frieling, "Andy Warhol" 90). For instance the *Do It Yourself* series, produced around 1962, invited audience members to finish paintings based on the paint-by-number hobby kits.

Although Joseph Beuys collaborated with Fluxus, and is sometimes described as post-Fluxus (Zimbardo, "Joseph Beuys" 130), his work remains distinct, but still very relevant for the debate on participatory art. Crucial to Beuys' work is his concept of the social plastic (also translated as social sculpture), which allows him to address social issues through artistic strategies. Beuys resists the 'traditional' formalistic and aesthetic definitions of art, and aims to dismantle society in order to build "a social organism as a work of art" (Beuys 125 – emphasis removed). In a lecture at the *Documenta 6* exhibition (which was telecast through satellite to more than 25 countries), Beuys said: "... such a notion of art would no longer refer exclusively to the specialists within the modern art world but extend to the whole work of humanity" (qtd. in Zimbardo, "Joseph Beuys" 130). The strategy Beuys uses is to try to convince each citizen to give form to life, at both the individual and the collective level. As Saper (23) remarks, this brings Beuys to combine an "extreme individualism" with a "collective internationalism." This also brings Beuys to his most often quoted statement: "Every human being is as artist" (125). One example is his *7000 Oaks* project, where for Beuys (and later his son) planted 7,000 trees in the German city of Kassel, where the *Documenta* exhibitions take place. Initiated at *Documenta 7* in 1982, Beuys and many volunteers intervened into the urban space of Kassel by planting these trees, each accompanied by a basalt stone. The project, with its biological, artistic, cultural, ecological and pedagogical components ended, one year after Beuys death, at *Documenta 8* in 1987 (see Beuys, Blume and Rappmann). Although Beuys charismatic figure and his position as an artistic celebrity (Frieling, "Towards Participation in Arts" 44) often complicated the participatory process in his own artworks, (some of) his work did allow for audience participation. One example here is the 100 day installation at *Documenta 5* (in 1972) called the *Bureau for Direct Democracy*. Here, Beuys installed an information office at *Documenta*, where he discussed the possibilities of direct democracy through the use of referenda (Beuys and Schwarze).

Finally, the neo-concretist movement, (co-)founded by South-Americans Lygia Clark and Hélio Oiticica, emphasised the need for participants to manipulate the artwork as way of understanding it (through their senses). The *Neo-Concrete Manifesto* from 1959 calls for the need to focus on more intuitive approaches to art and on natural subject matters, instead of following formulaic representational styles and reducing art to objects (Congdon and Hallmark 67). Clark's *Bichos* series used geometric metal constructions, inviting visitors to pick up, play with and/or stand on (ibid.); her *Dialogues* series aimed at creating dialogues between audience members, for instance by binding together their hands with a Möbius strip (Pellico, "Lygia Clark" 104). As Congdon and Hallmark (68) describe, the bodily senses played a key role in the participatory process: "She was concerned with activating participants' bodily senses along with their responses to those experiences." Oiticica also emphasised the bodily senses, for instance in his Whitechapel experiment where he "asked people to take off their shoes before entering large boxes filled with sand and straw or cabinlike structures with mattresses and blankets" (Pellico, "Hélio Oiticica" 107). Not surprisingly, dance became one of Oiticica media: he organised disruptive events with participants from Samba schools dressed in capes called *Parangolés* (see Braga). For Oiticica (106), the *Parangolé* "demands participation through dance," allowing for a "transformation of the 'total act of the self'." These transformations are seen as characteristic of what Oiticica calls 'environmental art':

> … being and indeed requiring the collaboration of various artists with differing ideas, solely concentrated on this general idea of a 'total participatory creation' – to which would be added works created through the anonymous participation of the spectators, who actually would be better described as 'participants' (108).

Frieling ("Towards Participation in Arts" 43) comments in a rather positive fashion on the work of the neo-concretist movement, writing that the organisation of "communal gatherings and discourses … pre-figured the idea of an open system that is constructed by participants-what we might call 'true' participation today."

In addition to these three arts movements, a similar emphasis on audience participation developed in the world of theatre. Again, this evolution was not new; already in his 1924 essay "Theater, Circus, Variety", László Moholy-Nagy had called for a new position for the audience:

> It is time to produce a kind of stage activity which will no longer permit the masses to be silent spectators, which will not only excite them inwardly but will let them take hold and participate-actually allow them to fuse with the action on the stage at the peak of cathartic ecstasy (25).

Also Bertolt Brecht (see Steinweg) had experimented with decreasing the audience / actor separation, for instance in his *Lehrstücke* project (which he abandoned, but was later reinvigorated the Brazilian director Augusto Boal's *Theatre of the Oppressed*, working with 'spect-actors' (Boal)). Especially in the 1960s and 1970s more structural changes in theatre theory led to the rearticulation of theatre as a text-based art to an open, playful and social event (Jackson and Lev-Aladgem 207). In so-called alternative and third theatre, audience participation implied "taking part in the play: dancing, playing a scene with the performers, engaging fellow spectators in conversation as part of the play, removing or exchanging clothing, or any of the many other kinds of physical involvement possible" (Schechner 73). One example here is the work of the Italian director Eugenio Barba and the Odin Teatret, who used a barter system in which the audience was invited to themselves (instead of paying for the performance) generate a performance: "A play is exchanged for songs and dances, a display of acrobatics for a demonstration of training exercises, a poem for a monologue, etc." (Watson 22).

After the heydays of participatory art in the 1960s and 1970s, it gradually became less popular, which has led Frieling ("Towards participation in arts" 45) to describe the 1980s as "a decade that avoided exploration of participatory social concepts." However, this does not imply that participatory art completely disappeared. As Manovich (57) points out, the first interactive computer installations already appeared in the 1980s. Before, the arts' fascination with media technology had provided important stimuli for audience participation, as for instance Herbert Schuhmacher's (Zimbardo, "telewissen") *Documenta der Leute* (People's Documenta) at *Documenta 5* (1972) shows. Schuhmacher's group used portable video equipment to interview passers-by and to play these interviews in a van outside the exhibition. A slightly later example was Jochen Gerz' (ZKM) 1980 video installation "Purple Cross for Absent Now", where he is seen in close-up on two video monitors, with a rubber rope around his neck. Visitors, who could only see Gerz and the other end of the rope, could then test the liveness of the situation by pulling the rope and tightening the noose. But especially the digital revolution stimulated participatory art, which became popular again in the 1990s.

Apart from a name change – interactive arts became a more popular denominator (see Dinkla) – there were a number of significant changes in the articulation of audience participation in this period, as also the societal context had structurally changes. As Nicolas Bourriaud (163) remarked in 1998, the "social utopias and revolutionary hopes [have] given way to day-to-day micro-utopias and mimetic strategies." In the 1990s, the emphasis was placed on interactive art that focussed on "the experience of the user as an act of communication, on the social space of the interface, and on the dynamics of interaction" (Penny 58). One of the key concepts of the era, Bourriaud's relational aesthetics, emphases human relations and context as starting point, where "[t]he status of the viewer alternates between that of a passive consumer, and that of a witness, an associate, a client, a guest, a co-producer and a protagonist" (168). Bourriaud (162) explicitly points to the responsibility of the artist ("for the symbolic models he is showing") and shies away from the concept of participation. Instead interaction and communication become emphasised:

> Their works bring into play modes of social exchange, interaction with the viewer inside the aesthetic experience he or she is offered, and processes of communication in their concrete dimensions as tools that can to be used to bring together individuals and human groups. (Ibid. 165)

Of course, participatory (or interactive) art in the 1990s and the 21st century is not exclusively new (or digital) media art. In 1997, the Interactive Arts Jury (107) of the Ars Electronica festival (one of the leading yearly media arts festivals) wrote that the 'usual' terms to evaluate interactive art were no longer adequate because of the inflation of the concept of interactivity. Moving away from a definition which links interaction to the digital – the 'usual' terms – they state that in interactive artworks the interaction "takes place between people, between people and machines, and between machines themselves." One of the many examples to illustrate the complexity of 21st century interactive art is Markus Kison's *Touched Echo*[2] (which was 'exhibited' at the *Brühlsche Terrasse* in Dresden from 2007 to 2009, but also at Ars Electronica in 2008). In this artwork, a small display invites the visitor to lean with his/her elbows on the railing (of the *Brühlsche Terrasse* or its substitute at Ars Electronica) and to cover the ears. If

2 See http://www.markuskison.de/touched_echo/.

these instructions are followed, sound is transferred from the metal of the railing into the body of the visitor and the visitor can hear the sounds of heavy planes and explosions, reminiscent of the bombing raid of 13 February 1945 which destroyed the city of Dresden. Other interactive art is characterised by the absence of technology. For instance in what Bhabha has called conversational art, and Finkelpearl dialogue-based public art, artists organise human interaction. Kester (77) for instance refers to the work of the Austrian arts collective Wochenklausur, who in 1994 brought politicians, journalists, sex workers and activists from the city of Zurich together in order to discuss drug policy. These dialogues eventually resulted in the establishment of a pension in Zurich, where drug-addicted sex workers could find a safe haven.

3. Moving participation into to the museum

The 1990s did not only witness a rediscovery of participatory / interactive art, also the institutions of display and conservation – the museum – became implicated in the debates about participation, as a series of museum theorists started to advocate a new museology or new museum theory. One foundational text was Vergo's reader, appropriately entitled *The New Museology*, in which he and a number of authors advocated a reconfiguration of our ways of looking at the museum. In his introduction, Vergo ("Introduction" 3) refers to the dissatisfaction with the 'old' museology, which focused too much on museum methods, and was not reflexive enough about the museum's purposes and identities. In the same introduction, Vergo also distanced himself from claiming ultimate novelty and exclusivity,[3] and mono-perspectivism. Within this diverse collection of articles, a number of authors (plead to) rethink the museum's relation to the visitor, and the power imbalances that characterise that relationship. For instance, Nick Merriman (167-168) – drawing heavily on Bourdieu's concept of distinction – concludes that "the action of museums in contemporary culture is to divide society into those who have the 'competence' to perceive museum visiting as a worthwhile leisure opportunity, and those who do not." Philip Wright (147) takes a similar position: "The present fiction in museums – that every visitor is equally motivated, equipped, and enabled 'to experience art directly'– should be abandoned. It is patronising, humiliating in practice, and inaccurate." Secondly, also the political nature of the museum and its functioning as a discursive machinery is thematised. Especially Greenhalgh's (96) chapter on international exhibitions offers a strong case, where he shows how these exhibitions "recognised the socio-political climate of their time and they responded to it."

3 See Halpin (1997) for a brief historical analysis of earlier museum (theory) reform projects.

In later publications on new museology / museum theory, this emphasis on representation, politics and power is deepened, and combined with a more explicit agenda for social and cultural change. Critiques on the elitism, exclusionary practices and mono-vocality of museums (Ross) form the basis of a museum reform project that aims for "the transformation of the museum from a site of worship and awe to one of discourse and critical reflection that is committed to examining unsettling histories with sensitivity to all parties." (Marstine 5) Secondly, also the emphasis in the inclusion of the museums' communities is continued, witness Marstine (ibid.) plea for a museum that "is transparent in its decision-making and willing to share power." Through this strong emphasis on inclusion and power, the notion of audience participation is brought into the debate again, for instance through the recognition that visitors and communities also have cultural expertise, as Halpin writes:

> the new or critical museology about which I am speaking might be a useful museology in service to a community, instead of the state and the élite. A museology practised by named, committed and creative professionals who know that people other than themselves are also cultural experts. (56)

Edited volumes like *Cultural Diversity. Developing Museum Audiences in Britain* (Hooper-Greenwill) and *Museums, Society, Inequality* (Sandell), but also monographs like *The Participatory Museum* (Simon) focus strongly on the importance of inclusionary practices, combined with the provision of series of examples. One example is Steve Hemming's chapter in the first edited volume, which has the (rather telling) title *Audience participation: working with local people at the Geffrye Museum, London*. In this chapter, Hemming discusses the exhibition *Chinese Homes: Chinese traditions in English homes*, which ran for three months in the Geffrye Museum in Hackney (London), in combination with the educational courses organised by the museum for different groups of people within the community. Without becoming over-celebratory: through the collaboration with a Chinese Community Centre, members of the Chinese community were involved in the construction of the *Chinese Homes* exhibition, by combining groups discussions on content (and access to preparatory meetings) with oral history approaches. In his process evaluation, Hemming points to the problems related to language, resources and time, but also emphasises the importance of audience participation:

> Involving the community in making decisions does take time, but also the will to make it happen. However, if the museum had tried to impose its own narrative on the exhibition without the consultation process, the results would have been disastrous. The chances are that the exhibition would have alienated the Chinese community and been a rather shallow attempt to portray their culture. (176)

4. Conclusion

This historical overview shows how cultural configurations and ideologies change, with avant-garde experiments with audience participation becoming significant cultural movements in the 1960s and 1970s, only to be abandoned in the 1980s, a decade which is sometimes referred to as the 'lost decade'. After this decade of silence, there was a new upsurge of participatory arts in the 1990s, but in a new structural setting, where the ambitions had been tempered. Interaction became the nodal point of the attempts to decrease the distance between artist and audience, resulting in a less radical reworking of the power balance between them. Still, within the same decade, also a series of cultural institutions became more sensitive towards these participatory issues, broadening the range of participatory theories and practices, but at the same time still softening their radical nature.

This trade-off between 'going deep' and 'going wide' brings us to the traditional interplay of resistance and incorporation, where radical ideologies are disarmed by their partial integration into mainstream systems and cultures. Arguably, we can see this dialectical process taking place over time, which illustrates the impact of discursive-ideological structures on how we see participation in a specific era, and what kind of intensity and balance we consider valuable and/or tolerable. It also shows that participation is not new, and that care should be taken not to fall into the trap of amnesia and to celebrate the present-day configuration yet again as the end of history.

At the same time care should also be taken not to romanticise the participatory experiments of the 1960s and 1970s. Moreover, it would be naive and even harmful to reduce the arts and museums to one specific democratic role, however important this role may be. But these experiments show that it is possible, within a plurality of artistic and exhibitionary practices, to structurally reconsider the relationship between art, artist and audience. Given the cultural status of the artwork and the artist, but also the position of museum professionals, this has never been an easy process, and there is much to learn from these periods of intense and artist-driven experimentation with participation, so that (a segment of) the arts and museum worlds can contribute to the continued democratisation of our cultural democracies.

References

Barthes, Roland. *Image, Music, Text.* Trans. Stephen Heath. London: Flamingo, 1984. Print.

Beuys, Joseph. "I Am Searching for Field Character." *Participation: Documents of Contemporary Art.* Ed. Claire Bishop. London: Whitechapel Gallery; Cambridge: MIT Press, 2006. 125-126. Print.

Beuys, Joseph. Bernhard Blume, and Rainer Rappmann. *Gespräche über Bäume.* Wangen: FIU-Verlag, 2006. Print.

Beuys, Joseph, and Dirk Schwarze. "Report of a Day's Proceedings at the Bureau for Direct Democracy, 1972." *Participation: Documents of Contemporary Art.* Ed. Claire Bishop. London: Whitechapel Gallery; Cambridge: MIT Press, 2006. 125-126. Print.

Bhabha, Homi K. "Conversational Art." *Conversations at the Castle: Changing Audiences and Contemporary Art.* Eds. Mary Jane Jacob and Michael Brenson. Cambridge: MIT Press, 1998. 38-47. Print.

Bishop, Claire. "Introduction: Viewers as Producers." *Participation: Documents of Contemporary Art.* Ed. Claire Bishop. London: Whitechapel Gallery; Cambridge, MA: MIT Press, 2006. 10-17. Print.

Boal, Augusto. *The Theatre of the Oppressed.* Trans. Charles A. McBride and Maria-Odilia Leal McBride. London: Pluto, 1979. Print.

Bourdieu, Pierre. *Distinction: A Social Critique of the Judgement of Taste.* Trans. Richard Nice. London: Routledge & Kegan Paul, 1984. Print.

Bourriaud, Nicolas. "Relational Aesthetics." *Participation: Documents of Contemporary Art.* Ed. Claire Bishop. London: Whitechapel Gallery; Cambridge: MIT Press, 2006. 160-171. Print.

Braga, Paula. "Hélio Oiticica and the Parangolés." *Third Text* 17.1 (2003): 43-52. Print.

Bruns, Axel. "'Anyone Can Edit': Understanding the Produser." Guest Lecture at SUNY, Buffalo / New School, NYC / Brown University, Providence / Temple University, Philadelphia. 22 Sep. 2005. Web. 25 June 2007. http://snurb.info/index.php?q=node/286.

Congdon, Kristin G., and Kara Kelley Hallmark. *Artists from Latin American Cultures: A Biographical Dictionary.* Westport: Greenwood Publishing Group, 2002. Print.

Debord, Guy. "Towards a Situationist International, 1957." *Participation: Documents of Contemporary Art.* Ed. Claire Bishop London: Whitechapel Gallery; Cambridge,MA: MIT Press, 2006. 96-101. Print.

Dinkla, Söke. "From Participation to Interaction: Towards the Origins of Interactive Art." *Clicking In. Hot Links to a Digital Culture.* Ed. Lynn Hershman Leeson. Seattle, WA: Bay Press, 1996. 279-290. Print.

Duchamp, Marcel. "The Creative Act." *Marcel Duchamp.* Ed. Robert Lebel. New York: Paragraphic Books, 1959. 77-78. Print.

Finkelpearl, Tom. "Five Dialogues on Dialogue-Based Public Art Projects." *Dialogues in Public Art.* Cambridge, MA: MIT Press, 2000. 270-275. Print.

Frieling, Rudolf. "Andy Warhol." *The Art of Participation: 1950 to Now*. Eds. Rudolf Frieling, Boris Groys, Robert Atkins and Lev Manovich. San Francisco: San Francisco Museum of Modern Art; New York; London: Thames & Hudson, 2008. 90-91. Print.

—. "Towards Participation in Arts." *The Art of Participation: 1950 to Now*. Eds. Rudolf Frieling, Boris Groys, Robert Atkins and Lev Manovich. San Francisco: San Francisco Museum of Modern Art; New York; London: Thames & Hudson, 2008. 32-48. Print.

Gerz, Jochen. "Purple Cross for Absent Now." *ZKM (Zentrum für Kunst und Medientechnologie Karlsruhe). Media Art Net.* 2004. Web. 9 July 2012. http://www.mediaartnet.org/works/purple-cross/.

Greenhalgh, Paul. "Education, Entertainment and Politics. Lessons from the Great International Exhibitions." *The New Museology*. Ed. Peter Vergo. London: Reaktion Books, 1989. 74-98. Print.

Groys, Boris. "A Genealogy of Participatory Art." *The Art of Participation: 1950 to Now*. Eds. Rudolf Frieling, Boris Groys, Robert Atkins and Lev Manovich. San Francisco: San Francisco Museum of Modern Art; New York; London: Thames & Hudson, 2008. 18-31. Print.

Halpin, Marjorie M. "'Play it Again, Sam': Reflections on a New Museology." *Museum International*, 49.2 (1997): 52-56. Print.

Hemming, Steve. "Audience Participation: Working with Local People at the Geffrye Museum, London." *Cultural Diversity: Developing Museum Audiences in Britain.* Ed. Eilean Hooper-Greenwill. London: Leicester University Press, 1997. 168-182. Print.

Hooper-Greenwill, Eilean, ed. *Cultural Diversity: Developing Museum Audiences in Britain.* London: Leicester University Press, 1997. Print.

Interactive Arts Jury Prix Ars Electronica. "Transcending the Categories. Statement of the Interactive Arts Jury." *Cyberarts: Internationales Kompendium Prix Ars Electronica / International Compendium Prix Ars Electronica* (The Prix Ars Electronica). Eds. Hannes Leopoldseder and Christine Schöpf. Wien; New York: Springer, 1997. 28-33. Print.

Jackson, Anthony, and Lev Shulamith-Aladgem. "Rethinking Audience Participation: Audiences in Alternative and Educational Theatre." *Theatrical Events: Borders, Dynamics, Frames*. Eds. Vicky Ann Cremona, Peter Eversmann, Hans van Maanen, Willmar Sauter and John Tulloch. Amsterdam; New York: Rodopi, 2004. 207-236. Print.

Kaprow, Allan. "Untitled Guidelines for Happenings." *Theories and Documents of Contemporary Art: A Sourcebook of Artists' Writings*. Eds. Kristine Stiles and Peter Selz. Berkeley; Los Angeles; London: University of California Press, 1996. 709-714. Print.

Kester, Grant. "Conversation Pieces. The Role of Dialogue in Socially-Engaged Art." *Theory in Contemporary Art since 1985*. Eds. Zoya Kucor and Simon Leung. Oxford: Blackwell, 2005. 76-88. Print.

Manovich, Lev. *The Language of New Media*. Cambridge, MA: MIT Press, 2001. Print.

Marstine, Janet. "Introduction." *New Museum Theory and Practice: An Introduction*. Ed. Janet Marstine. Malden, MA; Oxford, Carlton: Blackwell, 2006. 1-36. Print.

Melzer, Annabelle. *Dada and Surrealist Performance*. Baltimore; London: The Johns Hopkins University Press, 1976. Print.

Merriman, Nick. "Museum Visiting as a Cultural Phenomenon." *The New Museology*. Ed. Peter Vergo. London: Reaktion Books, 1989. 149-171. Print.

Moholy-Nagy, László. "Theater, Circus, Variety." *Multimedia, from Wagner to Virtual Reality*. Eds. Randall Packer and Ken Jordan. New York; London: W. W. Norton & Company, 2001. 16-26. Print.

Oiticica, Hélio. "Dance in My Experience." *Participation: Documents of Contemporary Art*. Ed. Claire Bishop. London: Whitechapel Gallery; Cambridge: MIT Press, 2006. 105-109. Print.

Pellico, Melissa. "Hélio Oiticica." *The Art of Participation: 1950 to Now*. Eds. Rudolf Frieling, Boris Groys, Robert Atkins and Lev Manovich. San Francisco: San Francisco Museum of Modern Art; New York; London: Thames & Hudson, 2008. 107. Print.

— "Lygia Clark." *The Art of Participation: 1950 to Now*. Eds. Rudolf Frieling, Boris Groys, Robert Atkins and Lev Manovich. San Francisco: San Francisco Museum of Modern Art; New York; London: Thames & Hudson, 2008. 104-106. Print.

Penny, Simon. "Consumer Culture and the Technological Imperative: The Artist in Dataspace." *Critical Issues in Electronic Media*. Ed. Simon Penny. Albany, NY: State University of New York Press, 1995. 47-73. Print.

Ross, Max. "Interpreting the New Museology." *Museum and Society* 2.2 (2004): 84-103. Web. 9 July 2012. http://www.le.ac.uk/museumstudies/m&s/ross.pdf.

Sandell, Richard, ed. *Museums, Society, Inequality*. London: Routledge, 2002. Print.

Saper, Craig J. *Networked Art*. Minneapolis, MN; London: University of Minnesota Press, 2001 Print.

Schechner, Richard. "Audience Participation." *The Drama Review: TDR* 15.3 (1971): 73-89. Print.

Simon, Nina. *The Participatory Museum*. Santa Cruz, CA: Museum 2.0. 2010. Web. 9 July 2012. http://www.participatorymuseum.org/.

Steinweg, Reiner. *Lehrstück und episches Theater: Brechts Theorie und die theaterpädagogische Praxis*. Frankfurt/Main: Brandes & Apsel, 1995. Print.

Vergo, Peter. "Introduction." *The New Museology*. Ed. Peter Vergo. London: Reaktion Books, 1989. 1-5. Print.

—, ed. *The New Museology*. London: Reaktion Books, 1989. Print.

Wagner, Richard. *The Art Work of the Future*. Trans. William Ashton Ellis. Whitefish: Kessinger Publishing, 2004. Print.

Watson, Ian. *Towards a Third Theatre: Eugenio Barba and the Odin Teatret*. London: Routledge, 1993. Print.

Wright, Philip. "The Quality of Visitors' Experiences in Art Museums." *The New Museology*. Ed. Peter Vergo. London: Reaktion Books, 1989. 119-148. Print.

Zimbardo, Tanya. "Allan Kaprow." *The Art of Participation: 1950 to Now*. Eds. Rudolf Frieling, Boris Groys, Robert Atkins and Lev Manovich. San Francisco: San Francisco Museum of Modern Art; New York; London: Thames & Hudson, 2008. 102-103. Print.

—. "Joseph Beuys." *The Art of Participation: 1950 to Now*. Eds. Rudolf Frieling, Boris Groys, Robert Atkins and Lev Manovich. San Francisco: San Francisco Museum of Modern Art; New York; London: Thames & Hudson, 2008. 130-133. Print.

—. "Nam June Paik." *The Art of Participation: 1950 to Now*. Eds. Rudolf Frieling, Boris Groys, Robert Atkins and Lev Manovich. San Francisco: San Francisco Museum of Modern Art; New York; London: Thames & Hudson, 2008. 98-100. Print.

—. "telewissen / Herbert Schuhmacher. Documenta der Leute (People's Documenta)." *The Art of Participation: 1950 to Now*. Eds. Rudolf Frieling, Boris Groys, Robert Atkins and Lev Manovich. San Francisco: San Francisco Museum of Modern Art; New York; London: Thames & Hudson, 2008. 143. Print.

Nico Carpentier is Associate Professor at the Communication Studies Department of the Vrije Universiteit Brussel (VUB – Free University of Brussels) and Lecturer at Charles University in Prague. His theoretical focus is on discourse theory, his research interests are situated in the relationship between media, journalism, politics and culture, especially towards social domains as war & conflict, ideology, participation and democracy.

Playful Public Connectivity and Heritage Institutions

Anne Kaun

Department for Culture and Communication
Södertörn University
Alfred Nobels allé 7
Stockholm, 14189
Sweden
Anne.kaun@sh.se

1. Introduction

Media theorist Neil Postman stressed in his notorious work *Amusing Ourselves to Death* that the Western industrial society oppresses itself by its addiction to amusement and entertainment, which is replacing deep elaboration and reasoning. For Postman the incarnation of entertainment culture was television, which cannot be taken seriously because it weakens rational inquiry. He assumes that amusement and entertainment on the one hand and serious engagement and learning on the other are entirely different. Postman stresses that television "know[s] about the value of providing the masses with amusements as a means of pacifying discontent" (141). Even if it is not the focus of Postman's inquiry, his basic assumption is that watching television is playful and therefore not serious, not rational and not important, but rather, watching television in counterproductive for society. Postman's reasoning found numerous updates discussing the ludification of culture, namely the increasing orientation towards entertainment. In the context of increasing competition between heritage institutions and other leisure activities, a similar critique is posed vis-à-vis entertaining and playful strategies for pleasure and attraction (Bruce), reinforcing Postman's contradiction between serious and playful engagement with history, art or technology. Focusing on the notion of play, several theorists from diverse fields, such as anthropology, sociology, ethnology and games studies, tried to overcome the separation between playfulness and seriousness. Psychological approaches and approaches from evolutionary theory for example focus on the functionality of play for individual development as well as the evolution of society. Current studies within game studies in particular focus on the transformative potential of play and games for learning. Rather than arguing for a certain usefulness of play and playfulness based on a functionalist perspective, this chapter aims to link playfulness with the concept of public connectivity and discusses it critically for the context of heritage institutions. The chapter will present a short overview of the theoretical discussion of the concept of play from a historical perspective, drawing on multiple disciplines, such as anthropology, sociology, philosophy, art history and museology, as well as media and communication studies. Based on the historical review, I will introduce the notion of playful public connectivity derived from the field of public sphere theory within media and communication studies and conclude with a discussion of the relevance of the concept for heritage institutions and critical museum theory. In that sense, I am not asking whether we are amusing ourselves to death but how amusement can be oriented towards an imagined, common ground.

2. The concept of play in history

Play is a contested concept that has been the subject of investigations in numerous disciplines, such as psychology, biology, anthropology, ethnography, sociology, and philosophy. In his article, *Evolving a Consilience of Play Definitions: Playfully*, Brian Sutton-Smith tried to identify common grounds for defining play across different disciplines to overcome the simplistic psychological essentialism dominating the field at the time. Whereas Imre started her investigation of the play concept in ancient Greece, where play was mainly understood in connection to learning and development, Sutton-Smith started his reasoning with an inventory of 20th century engagement with play. He stressed that a great commonality in the investigations of play during that period was a desire to show that play is useful (for example, by Huizinga and Turner) as a reaction to the dominating construction of play as a waste of time, for religious and economic reasons, over the previous 300 years. The rhetoric of play during the 20th century was, thus, mainly rehabilitating, foregrounding the contributions play has made to conventional socialisation. A typical characteristic of play stressed by that view is the sense of personal freedom and self-actualisation developed in the players resulting from this subjective experience. In the tradition of the Enlightenment, play was understood in terms of progress. Imre stressed that Kant's notion of "aesthetic judgment" started a phase of serious engagement with play, in which he understood play mainly as the free and unconstrained play of imagination. By linking play to a sort of freedom that is not completely unregulated, Kant inspired modern theories of play as important for progress and social development.

From an evolutionist perspective (Smith), play is mainly presented as skill preparation as argued, for example, by Karl Groos (in Imre). In that sense, play is understood as useful because it contributes to many kinds of development on moral, social, emotional and cognitive levels (Sutton-Smith). Children's play especially is a testing ground for social and moral rules, for example, because children learn through the pretense of playing a role (Mitchell 51). Similar to evolutionary psychology's development-orientated approach, the anthropological understanding of play focuses on play as a civilising force, mediating conflict and establishing communal identity and group hierarchy (Sutton-Smith). Play was presented as contributing to collective socialisation and social problem solving. Huizinga, for example, presented a new perspective in 1938 by identifying play as the foundation for all culture. According to Huizinga, play completely absorbs the players, includes uncertainty and is situated outside of the stream of ordinary life. The player is always completely conscious of the fact that it is "only" play and will, therefore, remain without consequences (Huizinga; Gordon and Esbjörn-Hargens). Gadamer, on the contrary, freed play from the dominating subjective perspective, arguing that in the context of art experience, play lies within the artwork itself. In his argument in *Truth and Method*, self-consciousness as a player is not a presupposition of play. Additionally, several other researchers emphasise the transformative character of play (Turner; Sutton-Smith; Imre). Victor Turner, in particular, stressed the playful character of liminal and liminoid rites, marking the transition, for example, between two periods in life. By analysing rites of passage, such as *Nkula, Wabwang'u* or *Wubina* of the Ndembu people in North-Western Zambia, Turner stressed the transformative character of playfulness beyond subjective freedom and creativity and shows tellingly the structured characteristics of play.

Silverstone critically reviewed Huizinga's play concept and stressed that Huizinga tends to misread the social significance of play and game because he placed them outside of the material world, excluding material interest from the very beginning. Furthermore, he criticised Huizinga's failure to develop a significant psychology of play. Therefore, Silverstone preferred the understanding of play put forward by Roger Caillois. Caillois pointed out that play involves freedom but within rule-governed limits. Hence, play is formalised, although by rules under conventions that suspend ordinary laws. Play is furthermore accompanied by a special awareness of second reality or of a free unreality: play is conducted with enjoyment. Play is therefore simultaneously both inside and outside everyday life, contained by holding structures of the ordinary but released from and legitimated by the ordinary as distinct and relatively unthreatening in their containment (Silverstone).

Both Imre and Sutton-Smith stressed that the modern understanding of play put a special emphasis on play's special role in society: "Modernity produced the world of play by threatening it and placing it under special protection from the increasingly encroaching world of work" (Imre 9). Instead of seeing play as wasteful, as in pre-modern times, play was valued as a symbol of modern freedom, as being active rather than passive, voluntary, indeterministic, intrinsically motivated, non-instrumental and creative (Sutton-Smith). Freud, for example, saw play as a possible escape from the burdens of the romantic idea of the family and as a space for creative freedom. At the same time, he did not oppose play to seriousness, but to reality. Similar to Roger Caillois, Sigmund Freud referred to the consciousness of the player about the act of playing, even if the player displays full emotional involvement. Furthermore, Freud stressed the creative moment of play and argued that creative writing and daydreaming continue children's play in adulthood.

Sutton-Smith criticised that these modern conceptions of play lack a linkage to Dionysian and irrational elements. In addition, he criticised the overstressing of the freedom of play because it is often obligatory and mandatory. Obligatory aspects of play include, for example, built-in rules and routines of play. Sutton-Smith argued that play compounds both voluntary and involuntary elements and asked if it would not be more fruitful to give up the idea of an unified function of play and instead develop an understanding of play as having multiple divergent frames of play (Sutton-Smith). Building on that assumption he developed the definition of play

> as a unique form of adaptive variability, instigates an imagined but equilibrial reality within which desequilibrial exigencies can be paradoxically simulated and give rise to the pleasurable effects of excitement and optimism, the genres of such play are humor, skill, pretence, fantasy, risk, contest, celebrations, all of which are selective simulations of paradoxical variability. (Sutton-Smith 253)

To amalgamate several of the partly ambiguous characteristics of play mentioned above and, at the same time, to include phenomena prior to play carrying play potentiality, I will instead examine playfulness, which enables an activity to become play. I prefer playfulness because it bears the potential to include a multitude of divergent forms of play being entirely different in character. Following Gordon and Esbjörn-Hargens, I understand playfulness as linked to "pleasure, freedom and the ability to bracket any experience within a larger context" (201). In addition, I understand play as distinct in time and space; formalised in one way or another; and linked to pleasure. However, I conceptualise playfulness as a continuum of varying degrees of phenomena, more or less full-filling these characteristics. In that sense, I incorporate Sutton-Smith's argument that play is ambiguous and, for example, includes both voluntary and involuntary moments. Additionally I conceptualise play as transformative in character (Turner), thus allowing public connectivity.

In summary, three main features of play have been identified as important for further analysis. Firstly, play is distinct in time and space. In that sense, the museum experience represents a particular, special experience, different from the stream of everyday life and routine. Secondly, play is formalised in one way or the other. There are norms and rules imbedded in the museum experience. Thirdly, play is linked to some form of amusement. In most cases, hopefully, the visit in the museum is more or less pleasurable in its outcome. The given definition is not to be understood as absolute but as a scale representing different degrees of playfulness of the distinct aspects of a museum visit.

3. Public connection and public connectivity

In the following section, I first define public connection and its components, as understood by Couldry et al., as a basic background to my concept of public connectivity, which will ultimately link to play. Thereby the chapter draws on broad discussions of public sphere theory, including research interested in the outcome of communication as practice, namely connectivity across media platforms, transgressing national borders, compressing time and space and, at the same time, establishing possibilities for common experiences (Harvey; Williams). One basic assumption within that field of research is that communicative action through rational discussion, potentially leading to deliberative consensus, is basic to democracy and the negotiation of shared values in a common culture as a process rather than a conclusion, which includes commonalities and differences, sharing and conflict (Williams; Habermas).

In their public connection project, Nick Couldry, Sonia Livingstone and Tim Markham started with a general discussion of the condition of contemporary democracy, asking if people in the UK still oriented towards a public beyond the purely private expressed in public connection. The authors defined public connection as an "orientation to a public world" (Couldry, Livingstone, Markham 3) and added that public connection is potentially "sustained by a convergence in the media people use" (ibid.). Applying this definition, the authors were interested to find whether people orient towards a public in times of media multiplicity and, as it is often argued, crisis of democracy. Thus, public is understood as a sphere where common concerns that reach beyond purely private matters (Couldry, Livingstone, Markham) are addressed. In the Habermasian understanding, the public sphere mediates between society and state, broadening to include cultural expressions of sociability.

Adopting main concerns of the public connection project, in the following I speak of connectivity rather than connection because connectivity points to a still ongoing process of building new connections instead of already established ones. At the same time, connectivity spans across a double-bind of connecting and disconnecting points. Kolb spoke of a theoretical duality, which includes the opposing categories being closely related and interdependent. Following Kolb, I define public connectivity as processes that link individuals and collectives to a discursively shared sphere of common concern by facilitating informational and social exchange. Public connectivity therefore encapsulates a latent potentiality of enhancing or hindering civic culture. Connectivity, in my understanding, is multi-faceted and constitutes sociality on different levels within society, the communication process being an important feature of civic culture. At any rate, only certain forms of connectivity are of interest here, namely those forms that bind the autonomous, private individual to a "public beyond the purely private" (Couldry, Livingstone, Markham 180). This form of connectivity I consequently will call public connectivity. Public connectivity is understood as precondition for engagement and participation in the (political or cultural) public.

In terms of connectivity, the centrality of large media institutions is not assumed. It is not necessarily only "the" media that structures orientation; it could also be museums or other heritage institutions. At the same time, public connectivity is embedded in the broad social field of practices and should therefore be analysed in that context without assuming media centrality or the centrality of any single institution. Following the argument of Couldry, Livingstone and Markham, it is possible to think of several forms of public connectivity. Those authors distinguished between media world connectors, public world connectors and multiple connectors, which can be thought of as an extension of Couldry's previous questioning of the myth of the mediated centre (Couldry; Couldry, Livingstone, Markham), in which he challenges the idea that media have a privileged relationship to a social centre. By addressing the myth of a social centre, where core values of (mainly national) societies are at stake, Couldry pointed out important questions of power to define what belongs to the social centre and what does not. Although public connectivity is thought of as an expression of sociality, it is not assumed that it links people to an essential social centre of core values to which people connect. Rather, those values are negotiated during the process of connectivity.

4. Playful public connectivity

In that context, playful public connectivity is distinguished from connectivity based on a perception of duty (Bennett, Wells and Rank; Zackariasson). Lance Bennett, Chris Wells and Allison Rank differentiated in a similar sense between dutiful and self-actualising citizenship, which is based on the assumption that youth today might approach civic culture and civic learning differently from their parents and teachers. This tendency finds its expression in a shift towards a declining sense of duty among young adults to engage and participate politically, while the engagement with issues that are fluid, changing, and lifestyle-oriented increased (Bennett, Wells and Rank). The underlying assumption of the duty-driven mode of connectivity is that to be a valuable part of society and contribute to sociality, one has to be informed about and potentially contribute to the political and cultural public sphere. Play, in contrast, refers to something that is intrinsically motivated and involves pleasure (Huizinga), which lies "outside of the world of duty and responsibility" (Stephenson). Examples for this playful, non-duty-oriented engagement with politics are visible in Jenkins's conception of "serious fun", which constituted a do-it-yourself political media culture made possible through new participatory media (Jenkins). Jenkins's examples of mastering political YouTube clips and photoshopping pictures of politicians into funny, sarcastic representations might be understood as expressions of voluntary forms of engagement, secluded from a daily routine and linked to pleasure. Playful new spaces of political negotiation seem to be taken by "ordinary" people formerly excluded from public discourse.

Elsewhere (Kaun), I presented empirical findings of playful public connectivity expressed in online diaries and in-depth interviews with Estonian students. There, playful public connectivity was mainly expressed as mediated entertainment reception that triggers connectivity with a shared public beyond the purely private. Humor (e.g., narration of political and socially critical jokes) and playful rituals (e.g., choir festivals) give but a few examples of what playful public connectivity might look like.

5. Playful public connectivity in museums

5.1 Heritage institutions as discursive public spaces

Heritage institutions, such as museums, tend to be perceived as neutral, objective grounds, as unmediated anchors in the past, educating the present. Although Marstine, referring to Gaskell, made clear that exhibited heritage objects have an afterlife, as well as a pre-life. Through decisions made by museum workers, e.g., the institution mission statement, architecture, exhibition display, accompanying texts, educational framing programs, and the logic of funding requirements, a narration is created, and heritage institutions become expressions of institutionalised constructions of collective memory. In the following, I focus on museums of all kinds, e.g., national museums, museums of ethnography, technology museums, and museums for modern and contemporary art. I do not intend to discuss other forms of heritage institutions, such as the UNESCO. Furthermore I am aware that 'museum' is, as in Hooper-Greenhills words, a "capacious concept" (3), with high variation and fluidity when it comes to the organisational forms and missions of museums. I take the position that the presented discussions apply to all forms of museums, though to different degrees. Traditional museum spaces construct a frequently national common, cultivated mainly through top down processes. Marstine asserted that museums frame the most basic assumptions about the past and selves within society. Through collecting, defining and revealing memory, museums construct a common both for the inside as well as for the outside of the institution. In that sense, heritage institutions could be understood as formal public spaces, potentially fostering public connectivity. The museum has long been seen as a place sheltering the authenticity of history with the special aura of the exhibits, including a potential for transcendent experience of the visitors (Marstine). However, museums are constantly renegotiating and redefining their own institutional identity, their mission and, therefore, their objects. In that sense, the museum as stable point of reference and timeless standard (Bruce) is illusionary. These renegotiations range from comprehensive corporate relaunches (e.g., Museum der bildenden Künste Leipzig,[1] which linked an image relaunch to the opening of a new museum building in December 2004) to rearrangements of exhibitions that give the museum a new character. In the 1990s, "new institutionalism" introduced, mainly in the context of contemporary art, a discussion about the negotiation processes within cultural institutions as such. One example is the European Kunsthalle project in Cologne, which made a constant renegotiation process as a cultural institution its focus point. The project saw itself as a "discursive platform without a physical location of its own" (European Society for Contemporary Art). In changing localities, the organisers arranged different exhibitions and projects to discuss and reflect the potentials of a "Kunsthalle" as a cultural model from a critical perspective.[2]

New museology and institutional critiques have critically addressed processes of mediation in heritage institutions and questioned possibilities for reflexivity and visitor participation. The seminal reader *The New Museology*, edited by Vergo, stressed that the relationship between visitors and institution has to be rethought in terms of the imbalances of power involved. Relying on the work of French philosopher Michel Foucault, Vergo's reader stressed the role of the museum as "discursive machinery", as Carpentier put it, and asked for a conception of museums as dialogical spaces, where visitors interact with the narration of the exhibition and negotiate their identity by displaying power relations between visitor and the institution. "Discursive machineries", such as museums, define the meaning of the exhibited objects in exchange with the exhibited and visitors, as Ushma Chauhan Jacobsen showed in her observational study of co-operational

1 http://www.mdbk.de/start.php4

2 http://www.eukunsthalle.com/70.0.html?&L=1

work between exhibitor and exhibited in the case of an exhibition of cultural artefacts from Swaziland in Denmark. Jacobsen clearly showed how intercultural differences are negotiated during the arrangement of the exhibition and focused on the inevitable reduction of the narrative complexity of exhibited cultural artefacts. What is told and what is not being told are her driving questions. In this process of developing the narration of an exhibition, the museum and its employees play a crucial role.

The museum or heritage institution could be understood as mediator of history representing the current value system. Therefore, it potentially serves as a point of reference for public connectivity, including forms of connection and disconnection.

5.2 Play and playfulness in heritage institutions

Similar to the initially presented discourse on television and amusement, play and playful elements in exhibitions are welcomed and condemned. Play and playfulness is often discussed in the context of transmitting content and engaging young visitors in particular with exhibitions (Bruce). The idea of play that contributes to learning processes has its roots in the ancient Greek word for play, which was closely linked to the word for education. In ancient Greek, play shared its root (paidia) with education (paideia) and children (paides). Furthermore, it was often mentioned in connection with pedagogy (paidagogia) (Imre). Therefore, it is not surprising that playfulness in exhibitions fosters involvement and engagement with the exhibits and the narration constructed in the exhibitions.

In the context of museums as educational institutions, play is often cited as a functional method for confronting and engaging visitors with knowledge in an involving manner (Hooper-Greenhill). Here a functionalist position cannot be dismissed. At the same time, it is clear that playful practices for knowledge transmission are restricted mainly to children and young adults.

Playfulness is also examined in what Chris Bruce described as the development of the importance of the fun factor in exhibitions. In the context of new museum professionalism, acknowledging that heritage institutions are in competition with other leisure activities, museums become places of entertainment rather than contemplation. There seems to be a broad agreement among museum theorists that museums today borrow exhaustively from the cinema and theme parks for the sake of the fun and spectacle factor that engages all the senses (Marstine). For example, one of the main attractions in the technological museum in Stockholm is the theme park-oriented testing ground for young visitors and a 4D cinema to explore the physical and technological wonders of the world.

By asserting the transformative potential of playful museum experience, Bruce critically questioned the development towards spectacle museum culture. In the need to attract the attention of the visitors, coinciding with a general decrease in public funding, many new museums and innovative exhibitions (e.g., Bruce's case study of the Experience Music Project in Seattle) focus on proactive engagement of visitors. Bruce did not neglect the beneficent educational motives of museums to apply entertaining and playful strategies but rather opens them for critical discussion. Often, the goals of what Bruce, following Hooper-Greenhill, terms as the post-museum are amalgamated in a dominating ideal of participatory, engaging exhibitions, which are supposed to herald a space for self-actualisation, particularly through playful features, as is stressed in 20th century theories regarding play. The driving argument for post-museums is the replacement of the edifying culture of traditional museums with active involvement and, therefore, empowerment of the visitors. Bruce critically addressed this ideal in his case study of the Experience Music Project in Seattle, showing that important features of the post-museum are missing from that museum, namely that the museum articulates its agenda, strategies and decision-making processes and continuously reflects them in exchange with the visitors (Marstine 19). In his case study, the visitors are not only witnesses to the spectacle but also become the spectacle, where any form of emancipatory engagement is dismissed.

In that sense, triggering playful public connectivity in museum exhibitions are two sides of one coin. On the one hand, one could argue for the transformative character and the positive potential of playfulness for fostering shared knowledge and involvement. On the other hand, playfulness could be interpreted as signs of maladaption to the crowd-pleasing methods of commercial entities. However, rather than taking one position for granted, I apply the concept of playful public connectivity in relation to critical museum theory and will potentially reach a differentiated idea of how museums could be participatory, engaging, inclusive and playful without simply being populist and consumer-oriented.

5.3 Critical museum theory and playful public connectivity

New museology, critical museum theory or new museum theory no longer conceptualise the museum as a neutral, objective institution but rather as one that frames and constructs. Protagonists within the paradigm of critical museum theory stress participation as a primary tool for opening the heritage institution as discursive public space but not only in terms of interactive exhibitions. They argue for a general openness for a vivid reflexive public discourse through transparency of decision-making and economic structures. In line with this, power relations are critically examined, not only internally but also in terms of external relations with, for example, visitors or involved communities. Employing playfulness and participation in exhibitions, potentially fostering playful public connectivity, cannot dissolve the asymmetries in those power relations. Part of the critical reasoning should, therefore, question whether playfulness is an expression of smart exhibition curation or maladaption to market requirements in addition to whether a reflexive moment on power relations within the rule definition of the game (Sutton-Smith) is needed. Who is defining playfulness? To what extent are the rules of the game open to definition and contestation from the outside, by communities and individual visitors? In that sense, a continuation of institutional criticism is needed, even if, and especially because, exhibitions are showing more and more engaging, playful and participatory moments. Even for definitions of participation as such, a meta-critique is necessary to contribute to the possibility of a critical museum visitor that reflects, participates and engages playfully, transformative with the displayed and non-displayed (Lindauer).

6. Conclusion

The chapter presented a historical overview of the diverse conceptualisations of the contested concept of play to conclude with a definition of playfulness as a varying continuum, ranging from voluntariness to involuntariness and secludedness to embeddedness in the everyday life, as well as from amusement to seriousness. Playfulness was furthermore linked to public connectivity as expressions of an orientation towards a shared common, fostered through and by playful activity. The article concluded by discussing the potentiality of playful public connectivity as a concept in the context of heritage institutions as discursive public spaces. In the light of the discourse on playful participation in museums, the article argued for a reflexive openness not only in terms of involvement in the exhibitions as such but also in the process of defining playfulness and its involving formalisations and rules, following a continuation of critical museum theory and the idea of the post-museum that does not end in a hypocritical, spectacle-centred, crowd-pleasing museum practice, which serves the purposes of commercial entities rather than enabling a critical visitor.

References

Bennett, Lance W., Chris Wells, and Allison Rank. "Young Citizens and Civic Learning: Two Paradigms of Citizenship in the Digital Age." *Citizenship Studies* 13.2 (2009): 105-120. E-Journal.

Bruce, Chris. "Spectacle and Democracy: Experience Music Project as a Post-Museum." *New Museum Theory and Practice. An Introduction*. Ed. Janet Marstine. Malden, MA; Oxford, Victoria: Blackwell Publishing, 2006. 129-151. Print.

Caillois, Roger. *Man, Play and Games*. 1961. Trans. Meyer Barash. Urbana; Chicago: University of Illinois Press, 2001. Print.

Carpentier, Nico. "A Short History of Cultural Participation." *Transforming Culture in the Digital Age*. Eds. Agnes Aljas et al. Tartu: Estonian National Museum, Estonian Literary Museum, University of Tartu, 2010. 11-19. E-Book.

Couldry, Nick. "Transvaluing Media Studies: Or, Beyond the Myth of the Mediated Centre." *Media and Cultural Theory*. Eds. James Curran and David Morley. London; New York: Routledge, 2006. 177-194. Print.

Couldry, Nick. *Listening Beyond the Echoes. Media, Ethics, and Agency in an Uncertain World.* Boulder; London: Paradigm Publishers, 2006. Print.

Couldry, Nick, Sonia Livingstone, and Tim Markham. *Media Consumption and Public Engagement: Beyond the Presumption of Attention.* Basingstoke; New York: Palgrave Macmillan, 2007. Print.

European Society for Contemporary Art. *European Kunsthalle.* Documents. Web. 9 Nov. 2010.

Freud, Sigmund. „Der Dichter und das Phantasieren." (The Poet and the Phantasy.) *Neue Revue. Halbmonatsschrift für das öffentliche Leben* 1 (1908): 716-724. Print.

Gadamer, Hans-Georg. *Truth and Method.* New York: Seabury Press, 1975. Print.

Gordon, Gwen, and Sean Esbjörn-Hargens. "Are We Having Fun Yet? An Exploration of the Transformative Power of Play." *Journal of Humanistic Psychology* 47.2 (2007): 198-222. Print.

Habermas, Jürgen. *The Structural Transformation of the Public Sphere: An Inquiry into a Category of Bourgeois Society*. Cambridge: Polity Press, 1989. Print.

Harvey, David. *The Condition of Postmodernity: An Enquiry into the Origins of Cultural Change*. Oxford: Blackwell, 1990. Print.

Hooper-Greenhill, Eilean. "Critical Pedagogy and the Post-museum". *Tingenes tale: Innspill til museologi. Bergen Museums skrifter* 12. Eds. Anders Johansen, Kari Gaarder Losnedahl and Hans-Jacob Ågotnes. Bergen: Bergen Museum, University of Bergen, 2002. 209-230. Print.

—. "Education, Communication and Interpretation: Towards a Critical Pedagogy in Museums." *The Educational Role of the Museum.* Ed. Eilean Hooper-Greenhill. London: Routledge, 1994. 3-27. Print.

Huizinga, Johan. *Homo Ludens: A Study of the Play-Element in Culture*. Boston, MA: Beacon Press, 1955. Print.

Imre, Anikó. *Identity Games. Globalization and the Transformation of Media Cultures in the New Europe*. Cambridge, MA; London: MIT Press, 2009. Print.

Jacobsen, Ushma Chauhan. "A World of Knowledge Asymmetries: The Professional Communication of Swazi Culture." Seminar Language Use and Cultural Identity. Helsinki 26 May 2010. Presentation.

Jenkins, Henry. *Convergence Culture: Where Old and New Media Collide*. New York: New York University Press, 2006. Print.

Kaun, Anne. "Playful Public Connectivity." *Transforming Culture in the Digital Age*. Eds. Agnes Aljas et al. Tartu: Estonian National Museum, Estonian Literary Museum, University of Tartu, 2010. 89-99. E-Book.

Kolb, Darl G. "Exploring the Metaphor of Connectivity: Attributes, Dimensions and Duality." *Organization Studies* 29.1 (2008): 127-144. Print.

Lindauer, Margaret. "The Critical Museum Visitor." *New Museum Theory and Practice. An Introduction*. Ed. Janet Marstine. Malden, MA; Oxford; Victoria: Blackwell Publishing, 2006. 203-225. Print.

Marstine, Janet. "Introduction." *New Museum Theory and Practice. An Introduction*. Ed. Janet Marstine. Malden, MA; Oxford; Victoria: Blackwell Publishing, 2006. 1-36. Print.

Mitchell, Robert W. "Pretense in Animals: The Continuing Relevance of Children's Pretense." *Play and Development. Evolutionary, Sociocultural, and Functional Perspectives*. Eds. Artin Göncü and Suzanne Gaskins. New York; London: Lawrence Erlbaum Associates, 2006. 51-76. Print.

Postman, Neil, and Andrew Postman. *Amusing Ourselves to Death: Public Discourse in the Age of Show Business*. 20th anniversary ed. New York, NY: Penguin Books, 2006. Print.

Silverstone, Roger. "Rhetoric, Play, Performance. Revisiting a Study of the Making of a BBC Documentary." *Television and Common Knowledge*. Ed. Jostein Gripsrud. London: Routledge, 1999. 71-90. Print.

Smith, Peter K. "Evolutionary Foundations and Functions of Play: An Overview." *Play and Development. Evolutionary, Sociocultural, and Functional Perspectives*. Eds. Artin Göncü and Suzanne Gaskins. New York; London: Lawrence Erlbaum Associates, 2006. 21-50. Print.

Stephenson, William. *The Play Theory of Mass Communication*. Chicago: University of Chicago Press, 1967. Print.

Sutton-Smith, Brian. "Evolving a Consilience of Play Definitions: Playfully." *Play & Culture Studies*. Vol. 2. *Play Context Revisited*. Ed. Stuart Reifel. Stamford, CT: Ablex Publishing Corporation, 1999. 239-256. Print.

—. *The Ambiguity of Play*. Cambridge, MA; London: Harvard University Press, 1997. Print.

Turner, Victor. *From Ritual to Theatre. The Human Seriousness of Play*. New York: Performing Arts Journal Publications, 1982. Print.

—. *The Ritual Process: Structure and Anti-Structure*. New Brunswick, NJ; London: Aldine Transaction, 2008. Print.

Williams, Raymond. *Culture and Society: 1780-1950*. New York: Columbia University Press, 1983. Print.

Zackariasson, Maria. "Lust eller plikt? Motiv för politiskt engagemang bland utomparlamentariskt aktiva unga." (Pleasure or Duty? Motivs for Political Engagement among Non-Parliamentarian Active Youth.) *Tidsskrift för Ungfomsforskning* 8.1 (2008): 31-48. Print.

—. *Viljan att förändra världen: Politiskt engagemang hos unga i den globala rättviserörelsen*. (The Will to Change the World: Political Engagement among Young People in the Global Justice Movement.) Umeå: Boréa Bokförlag, 2006. Print.

Anne Kaun, PhD, works as Researcher and part-time Lecturer at the School of Culture and Communication at Södertörn University, Stockholm. During her PhD studies she has been enrolled in BEEGS – Baltic and East European Graduate School. She has published on solicited, open-ended online diaries and civic experiences of young Estonians.

Solitude in Cyberspace

Piret Viires

Professor
Tallinn University
Estonian Literary Museum
Narva mnt 25
Tallinn, 10120
viires@eki.ee

Virve Sarapik

Professor
Estonian Academy of Arts
Estonian Literary Museum
Roosikrantsi 6
Tallinn, 10119
virve@eki.ee

1. Introduction

The relationship between literature and cyberspace has attracted the attention of literary researchers since the 1980s, when the first works of digital or electronic literature appeared. Digital literature, broadly, is literature that has been created by and is read on the computer. The Electronic Literature Organization has suggested the following definition of electronic literature: "work with an important literary aspect that takes advantage of the capabilities and contexts provided by the stand-alone or networked computer" (see Hayles 3). N. Katherine Hayles offers the following explanation: "Electronic literature, generally considered to exclude print literature that has been digitized, is by contrast "digital born," a first-generation digital object created on a computer and (usually) meant to be read on a computer" (3). Studies of digital literature have also been closely connected with the analysis of hypertext literature and later the cybertext theory (cf. Aarseth).

Analyses of digital literature and cybertexts mainly derive from the keywords of increasing collectivism: shared authorship, reader-viewer interaction, their active participation in creating text etc. (cf. e.g. Hayles; Koskimaa).

However, the current article does not tackle the aspect of the collectivism of digital literature and instead focuses on the opposite phenomenon: the essence of individualism in the process of digital text creation, solitude. We next analyse solitude in cyberspace from two perspectives: first, examining the solitude of the creator; and then examining the solitude of creative work. Within the framework of these general theoretical problems, the second part of the article analyses the processes that have occurred in Estonian literature in order to point out the common features and differences.

The term solitude naturally also refers to the romantic principle of creative work and is associated with such aesthetic concepts of art as authorial self-expression and art for art's sake. The current paper disregards this aspect, as well as the emotional and psychological connotations of the word. Solitude is here regarded as a technical term, indicating the number of different agents in the creative process. This primarily means whether the text can be associated with one author and his/her intention, or whether authorship is distributed between several people who have participated in the creation of the text; and if and how much texts presume activity on behalf of the reader. In the case of collective authorship, we can, in turn, distinguish synchronic creative joint work or participation of different people in various stages and functions of text creation (e.g. the programmer, sound technician, or designer of the visual part). The approach to solitude is therefore somewhat conservative, as it takes into account the relationships between direct and identifiable agents in the creative process and not, for example, the subliminal intertextual processes.

2. Solitude and the author's responsibility

We can claim that when writers write their texts they are usually on their own. A text is born in the writer's head and he or she needs some kind of form to present it. When the form of literature was mostly what was recorded on paper, we could say that the author formalised the text in solitude – writing alone on pieces of paper. Only when handing in the manuscript were other participants added to the completion process of the literary work: the editor, designer and printer. Each era has naturally also had authors keen to discuss their incomplete texts with friends, and introduce half-completed excerpts to others, as well as the more indirect influences of the surrounding people, era, etc. Such authors are thus, to some extent, influenced by others, although this depends on the author's character rather than on the form of the text.

If we expand the historical dimension of the problem, two aspects can be added: peculiarities of oral and collective creative work (folklore), and the anonymity of the writer preceding modernist authorship (cf. e.g. Kris, Kurz 3; Genette 42, etc.). During the Renaissance, the position of the creator and the status of art shifted. An artist was no longer an anonymous intermediary of God's text, but an individual subject who had a significant life story and a divine gift to be himself the creator of masterpieces (e.g. Kris, Kurz 13-60).

In his essay "What is an Author?" Michel Foucault points out that the author emerged on the threshold of the modern era and modernity led not only to his independence as a creator, but also to social responsibility. The author was punishable if the text was illegal, blasphemous or in opposition to some other accepted norms in society. This aspect has accompanied the author-text relationship to this day, pulverised and blended in variations of Internet anonymity, personality and collectivism.

Following this line of thinking, we can claim that the modern author's position (public name) not only glorifies the author as a divine creator (e.g. Kris, Kurz 38, etc.) but, in fact, forces him or her into solitude together with the work created. The author becomes solely responsible the text when adding his or her name to it. A person generally has a unique name and this singularity separates the author's person and his or her creation from other people and texts. Responsibility and the author's name simultaneously denote appreciating both personality and God, but also the creator's burden. The text survives in the textual world, living its own life and, being sealed with the author's name, it will never release the author as a person.

Although the ideas expressed in Roland Barthes' "The Death of the Author", which preceded Foucault's essay, are generally considered more radical than Foucault's – such as the author giving up the creator's position to the text – the opposite can also be claimed. It is Barthes who tried to liberate authors from the burden of social responsibility, thus keeping them free for creative work. The author lacks control over the fate of the completed text, but he also lacks responsibility. The text no longer accompanies the author forever as a completed and perhaps slightly shameful act. The author's function as suggested by Foucault is to be an agent who categorises and appreciates texts, separated from the author as a human person, and to be a creator controlling the meaning of the text. This does not come before the text, does not condition it, but always exists in the text to varying degrees. On the other hand, authors have a constant responsibility, and their function continues to be "linked to the juridical and institutional system that encompasses, determines, and articulates the universe of discourses" (Foucault 216). To get rid of this function, in Foucault's opinion, is only a hypothetical future possibility (Foucault 391).

In subsequent years, the ideas of both Barthes and Foucault have been questioned from different starting points. For example, when we consider the unavoidable intertextual relationships of texts (nothing is born out of nothing), the author of every era as an individual creator, inevitably becomes vague (cf. e.g.

Burke). On the other hand – from the point of view of a writing individual – at the beginning of modern times, the heroic role of the creator and hence of the 'owner' of the text, acquired in new social conditions, still seemed so attractive that abandoning it kept evoking the author's intuitive opposition.

We can thus generally claim that the work of a post-typography author, i.e. the author in the modernist era, is born in solitude. This solitude is an inevitable side effect of the emergence of a renaissance creative person, uniting the positive element of emerging individualism with the torments of social responsibility and commercial relationships.

3. Solitude and digital literature

We can, however, assume that these relationships change significantly in the case of digital literature where the production is not a printed book, but a cyber- or hypertext, and the following focuses on the causes and essence of these changes.

The changes are caused by at least two circles of reasons. The first naturally involves the peculiarities of creating digital literature (more about this below). The second involves the position and functioning of digital literature in society as a whole, including: (1) its relationships with traditional mainstream printed literature, (2) its impact nationally or beyond the national, and (3) its relationships with other types of texts (primarily mass media) and art.

In addition to the author of the text, cybertexts and hypertexts require active co-authors: programmers, designers and others. Creating a cybertext is therefore basically a collective act, although there are of course exceptions. An electronic writer or poet can thus be technologically skilled and also familiar with web design, or might use special programmes for writing hypertext literature (e.g. Storyspace). However, an ordinary writer who wants to experiment with hypertext literature does not, on the whole, have sufficient technological knowledge and needs assistance. Relying on the work "The Functional Point of View" (1997) by Philippe Bootz, the electronic poet Jim Rosenberg has compared the author of a cybertext with a filmmaker (83). "The cybertext author creates a texte-auteur, a kind of scenario by which the cybertext is assembled. The cybertext author becomes like a filmmaker, having to extrapolate in his mind how the final piece will work" (ibid.). And just like a filmmaker, a cybertext author needs a team who realises his ideas. The author of a cybertext is no longer the sole and unique creator.

This kind of creative method can be compared with the production process of films, rather than the old traditional literary models. It should also be borne in mind that the Auteur theory in film criticism, connected with the French New Wave cinema, developed fully as late as the 1950s-1960s. The theory emphasised the director's creative role regarding the production process, screenwriter, producer and actors (see cf. Sarris; Naremore 10-16; Stam 83-92). The theory, greatly influenced by existentialism, did not favour collectivism but, quite to the contrary, tried to shake off this pressure and fight for the rights of the modernist individual creator. Indeed, the ideal was an individually creating writer: the mechanisms and production of films were supposed to become a pencil held by the director (Astruc). Although in the subsequent structuralist and neo-formalist film criticism the prestige of the Auteur theory waned, some of its attitudes survived. An authorial film, i.e. creative work clearly connected with an individual, today enjoys a higher status, and the issues of individual style have not disappeared (cf. also Andrew 77-85; Naremore 20-23).

Some similarities can thus be detected between digital literature and the practices of film production.

4. Digimodernist turn

Alan Kirby has introduced the concept of digimodernism, which marks the cultural stage connected with the spread of Web 2.0 in the early 21st century (cf. Kirby, "Digimodernism"). This can also be associated with Henry Jenkins's concept of "participatory culture" and the democratisation of culture (cf. e.g. Jenkins). Kirby finds that, since the mid-1990s, digimodernism has replaced postmodernism and has "established itself as the twenty-first century's new cultural paradigm" ("Digimodernism" 1). Digimodernism is directly connected with technological development and "identifies as the critical event in contemporary culture the profound and shattering encounter between computerisation and the text. Its most recognisable form is a new kind of digitised textuality – onward, haphazard and evanescent – that disrupts traditional ideas about authorship and reading, and is found on Web 2.0, a range of applications and in the video game" (Kirby, "Successor States to an Empire in Free Fall"). In short, digimodernism is a phenomenon that is connected with, for example, the blogosphere, Wikipedia and Facebook – everything in which there is lively interaction between authors, readers and users in general. The "digimodernist turn" in the form of blogs, wikis, Facebook and Twitter, has also brought about a change for authors of digital literature. The technological simplicity of the software of the new web platforms, and especially the simplicity of blog software, means that the authors no longer need any urgent technical assistance. Although Alan Kirby himself has expressed doubt about whether digimodernist literature actually exists ("Another Interview I Gave"), we can regard, for example, the numerous literary blogs as being digimodernism. Most blogs have been seen as descriptions of everyday life, writings focusing on a specific topic or as alternative journalism. Apart from the autobiographical aspect of blogs and the form of life-writing, the blog researcher

Jill Walker Rettberg, for instance, has noted the fictional essence of some blogs and analysed the blogs' narrative structures (111-126). Rettberg also refers to similarities between blogs and hypertext literature: linked text fragments, open endings, and the possibility of disrupting the narrative. In addition, Rettberg points out the intentional fictionality of some blogs, and how they play on the borders of reality and fiction (e.g. Kaycee Nicole, lonelygirl15). There are blogs that are consciously written as fictional narratives, thus most resembling the online hypernovels (Rettberg 121-126). The interaction between authors and readers in the blog world is already familiar from the earlier hypertext literature. Espen Aarseth, for instance, has explained the changed roles of authors and readers, saying that while generally a reader's role is limited to interpretation, in terms of digital literature a reader or a user serves four functions. A reader has an active role – in addition to interpretation he navigates, configures and writes (Aarseth 64).

Connections of the blogosphere and literature have also been noted by other authors besides Jill Walker Rettberg. Aimée Morrison has written that the community of literary studies has begun to examine the links between diary-style blogging and other forms of life writing, while links between blogs and prior internet forms are also seen (such as "online diaries" in simple html that began appearing around 1995) as well as print antecedents (paper journals, diaries and scrapbooks).

Digimodernist literary blogs might for example be the blogs of professional writers, in which autobiographical stories are presented in a polished literary form. Or they might be blogs that mix the factual and the fictional, blogs in which the author knowingly plays around at the borders of fabrication and reality, or blogs in which the author tells the readers in advance that they are reading fiction. There are also blogs presenting some literary narrative and blogs publishing poetry. The concept of the writer in the digimodernist era has undergone a shift; in addition to professional literature, the Internet is full of everyman literature, as it were. However, it is possible to claim that certain blogs constitute literary phenomena, namely those that clearly display their literary content and possess at least a hint of quality.

When we turn to the topic of solitude, it should be pointed out that the technological simplicity of the digimodernist turn brings forth the problem of the author's solitude – the author is once again formalising his or her work alone, just as with printed books.

At the same time, we can speculate that the solitude of a creative work in the cyberworld disappears. After publishing a book in print, the author loses control over it. In other words the work is left on its own or, to be precise, the text begins living its own life independently of the author: it is finished, final and the author usually can no longer change it.

In the cyberworld, on the other hand, connections in various forms between the author, the work and the reader are retained. The most significant difference, however, is the retaining of the *author's controlling function. The author reacts to comments, can close down his blog at any time and can edit the already published texts.*

It might thus seem paradoxical that in the printed world both the author and the work are solitary, whereas in the cyberworld the solitude of creative work vanishes, because cybertexts, as well as blogs, generally need feedback from readers and interaction between authors and readers. At the same time, the author's solitude in cyberspace is twofold – creating cybertexts often requires some assistance, whereas digimodernist blog literature can be produced in solitude, independently.

5. *Staircase* and *Sonnet Machine*

Very few cybertexts in Estonian digital literature have been produced as teamwork, with technical assistance. One of the first was Hasso Krull's hypertextual poem *Trepp* (*Staircase*, cf. Krull). Krull, a poet always eager to experiment and familiar with post-structural theories, tested the possibilities of creating a hypertext at an early phase when the Internet had just arrived in Estonia. This is a classic hypertext poem and is essentially simple. The poem has a fixed beginning, and navigating along links makes it possible to read the poem differently. Through the selection of a certain path, the poem reaches the ending chosen by the author. Navigating by other paths, however, readers might find themselves in a circle without an exit. The poem is of excellent literary quality and, with its various links, offers different poetic impressions. The links are emphasised by different colours, and the beginning and ending of the poem are marked by a picture.

This hypertextual poem is a typical case where the poet tries to experiment with new technological possibilities, but needs the help of a computer expert to realise the idea. The web design and the linking of the hypertext were organised by Krull's technical assistant. This was Krull's only attempt at hypertext poetry, and after that he returned to traditional forms of poetry.

One of the grandest team projects was the *Sonnet Machine*, initiated by the poet, translator and cultural theoretician Märt Väljataga in 2000. He was inspired by Raymond Queneau's work *Cent mille milliards de poèmes* (1961), although Väljataga developed his project *Sada tuhat miljardit millenniumisonetti* (One Hundred Thousand Billion Millennium Sonnets) to the extreme. By combining different strips in a book of strips (cf. Väljataga), it was indeed possible to form 100 000 000 000 000 sonnets. In addition, an Internet text generator was created in 2000, which allowed the user to combine lines in different positions and could thus produce $(20 \times 19)^7 \times 7$ = 5 766 545 362 636 800 000 000 English sonnets. It could also produce a huge number of Italian sonnets and 14-verse poems with other rhyme schemes. In addition to the book and the web version, Väljataga and his assistants constructed an enormous sonnet machine and displayed it in an art gallery. The machine was electromechanical and by operating it manually it was possible to create millions of sonnets. Väljataga's original idea of taking Raymond Queneau's idea further, thus resulted in three solutions: on paper, on the Internet and in the huge sonnet machine. What should be pointed out here is that Väljataga was not able to realise a single one of his ideas on his own. The designer helped with the book, the Internet text generator was mainly provided by the programmer Hannu Krosing, and the co-authors of the monster artefact were the artist Lennart Mänd and the composer Rauno Remme (see also Kesküla). Unfortunately, Väljataga did not experiment any further: he produced no more Internet projects and built no more new mechanical poetry machines. The most curious and regrettable fact about the whole project is that only the printed book still exists today. The text generator disappeared from the net after the crash of the server and the sonnet machine installation was dismantled. Only a few pictures have survived of the installation.

Considering the technological experimentation of Estonian writers, these two examples are exceptional rather than usual. There have been only a few other smaller projects (e.g. Tambet Tamme's project *The Weather Station Never Lies*, and Lemmit Kaplinski and Jaak Tomberg's *Prepare*).

6. The Estonian writer and digimodernism

Estonian authors have been reluctant to try computer-technological experiments, despite the general fascination with technology in Estonian society. However, the digimodernist change is altering the situation. New technological platforms are widespread and easy to use, which has resulted in the fact that, all of a sudden, Estonian writers seem to have overcome their aversion to technology and are eagerly using the possibilities offered by Web 2.0. Many Estonian writers are also active bloggers and Facebook users (e.g. Kivisildnik, Olavi Ruitlane, Aare Pilv, Tõnu Õnnepalu and several younger authors). The writers' blogs include all aspects of the above-described literary blogs: there are autobiographical descriptions, the authors publicly declare that they are producing fiction, they publish poetry, and they play on the borders of the factual and the fictional (e.g. the blog of Olavi Ruitlane, who has passed autobiographical material off as fiction). Several writers express their opinions on literature and literary life. And quite a few have published their blogs in book format. Several literary discussions have also moved to Facebook, where the authors often refer to their blogs, where they publish their new work or discuss some literary issue at greater depth than Facebook's limited space allows.

In summary, Estonian writers who were earlier afraid of technology, have become very keen on it in the digimodernist world.

What could have caused this change? Why have Estonian writers not been eager to conduct elaborate computer-technological experiments that require co-operation with different specialists, but are ready to actively blog and use technological platforms, which are easy to use and thus do not demand any assistance?

7. The Estonian author as a solitary author

We would like to propose a few possible solutions. Firstly, there is a wish to be independent. Estonian writers want to be solitary in cyberspace, write on their own, be fully responsible and not participate in teamwork, in a collective act of creative work. This could be explained by the well-worn thesis about the smallness of Estonian literature, which limits the experiments and extremes outside the mainstream. The reason might also be the general conservative attitude towards all cultural manifestations in the otherwise technology-keen Estonian society.

Of course, we can argue that there are authors of digital literature who are technologically skilful and are familiar with programming, and therefore we can speak in these cases only of one, single author, not about cooperation between different agents. However, such a person unites various agents or roles in himself (as does the writer who illustrates his own works and the poet setting his poem to music).

If we look at parallels in other areas, we see that, for example, the Estonian visual art in the first half of the 1990s quite easily accepted video and installation practices that offered technological opportunities; as well as accepting opportunities offered by the Internet and digital art. Numerous examples can also be found in music, and syntheses of music and pictorial art. From there onwards, different collective practices, the employment of technical assistance and the synthesised forms of various means of expression have become an organic part of visual art (see also Kelomees, "The Material and Technological in the Art of the 1990s", "Jagatud autorsus: kunstniku hajumine elektroonilistel väljadel"; Sarapik, "Estonian Art of the 1990s and the Postsemiotic Turn", "Semi-Closed Openness: Transformations of Estonian Art and 'Meta-Art' in the 1990s"). Traditional art forms – prints, oil paintings and sculpture – have naturally not disappeared either. At least at the beginning, the processes of redeveloping the art sphere were rather painful (e.g. representatives of more traditional fields often felt pushed aside). At the same time, we could point out synchronic tendencies with literature if we expand the scope of observation to the entire art discourse. Until recently, art reception followed the traditional publishing channels: dailies, cultural periodicals and other publications. Art discussions in printed media in the early 1990s were extremely lively, whereas today art criticism in the daily papers has pretty much vanished. Art discussions have not made use of the opportunities offered by the Internet and there has only been one animated art discussion list, called Latera, and this lasted only a few years (2000-2004, Sarapik, "Semi-closed Openness: Transformations of Estonian Art and 'Meta-art' in the 1990s" 72). Changes have occurred only in the last few years, due to the emergence of several active art blogs (primarily the collective blog Artishok, http://artishok.blogspot.com/, active since 2006). In addition, Facebook is becoming increasingly prominent. At the moment, we could say that art discussion among professionals has largely moved to the Internet, and the print media has failed to offer significant alternatives.

Literature in Estonia is therefore, in a way, continuing the tradition of the modernist author who does not wish to give up the position of being an individual author. Digimodernist technological simplicity has indeed made possible the organic transfer from printed text to digital literature, which does not endanger the authorial position. This seems to be confirmed by the fact that even writers who are active in the blogosphere, or on Facebook or Twitter are still primarily interested in publishing their messages, and not so much in exchanging ideas, dialogues, etc., which might cause the message to be changed or transformed into collective work.

8. Conclusion

In conclusion, we could say that an Estonian writer is essentially a solitary author.

Cyberculture, however, further emphasises the paranoid position of a 'dead' author in terms of completed text. The printed text can be interpreted, although its original form can no longer be altered, whereas alteration is often possible in the digital sphere. A blog text can be changed or locked, and cyberworks can be deleted. This is mostly done by the controlling, hesitating and later regretting author.

At the same time, the public cyberspace is full of endless fragments about people, of which these people might not even be aware - they certainly cannot control the existence of these fragments. It seems that digimodernism eliminates the creative solitude and the responsibility of the author (e.g. the author can correct the texts later), but at the same time it is sometimes imaginary: collective authorship, participation in the author-reader relationship, and as well as seemingly abandoning responsibility. Collective forms of creating have at least initially maintained the traditional form: groups, creative communes, professional associations and circles of friends. The most vigorous communication in cyberspace occurs between people who already know one another. And although Foucault finished his essay with the optimistic hope that with the changing society, the author's function will disappear and be replaced by the "anonymity of murmur" of discourses (Foucault 222), cyberculture is only just starting down that path.

Acknowledgements

The article was written with the support of Estonian Science Foundation grant no. ETF7679 "Participatory Culture in Cyberspace: Literature and its Borders" and targeted financed research project no. SF0030054s08 "Rhetorical Patterns of Mimesis and Estonian Textual Culture".

References

Aarseth, Espen J. *Cybertext. Perspectives on Ergodic Literature*. Baltimore, MD; London: The Johns Hopkins University Press, 1997. Print.

Andrew, Dudley. "The Unauthorized Auteur Today." *Film Theory Goes to the Movies: Cultural Analysis of Contemporary Film*. Eds. Jim Collins, Hilary Rodner and Ava Preacher Collins. New York; London: Routledge, 2003. 77-85. Print.

Astruc, Alexandre. "The Birth of a New Avant-Garde: La Caméra-Stylo." *Film and Literature: An Introduction and Reader*. Ed. Timothy Corrigan. 1948. Upper Saddle River, NJ: Prentice Hall, 1999. 158-162. Print.

Barthes, Roland. "The Death of the Author." *Image-Music-Text*. Trans. Stephen Heath. 1968. New York: Hill and Wang, 1977. 142-148. Print.

Bootz, Philippe. "Le point de vue fonctionnel: point de vue tragique et programme pilote." *alire 10 / DOC(K)S Series* 3.13/14/15/16 (1997). 28-47. Print.

Burke, Seán. *The Death and Return of the Author. Criticism and Subjectivity in Barthes, Foucault, and Derrida*. 1992. Edinburgh: Edinburgh University Press, 1998. Print.

Foucault, Michel. "What Is an Author?" *Aesthetics, Method, and Epistemology. Essential Works of Foucault, 1954-1984*. 2. Ed. James D. Faubion. 1969. New York: The New Press, 1998. 205-222. Print.

Genette, Gérard. *Paratexts: Thresholds of Interpretation*. 1987. Cambridge; New York: Cambridge University Press, 1997. Print.

Hayles, N. Katherine. *Electronic Literature: New Horizons for the Literary*. Notre Dame, IN: University of Notre Dame Press, 2008. Print.

Jenkins, Henry. *Fans, Bloggers, and Gamers. Exploring Participatory Culture*. New York; London: New York University Press, 2006. Print.

Kaplinski, Lemmit, and Jaak Tomberg. *Prepare*. 1999, 2002. Web. 11 Apr. 2010. http://lemmit.kaplinski.com/prepare/.

Kelomees, Raivo. "The Material and the Technological in the Art of the 1990s". *Ülbed üheksakümnendad. Probleemid, teemad ja tähendused 1990. aastate eesti kunstis / Nosy Nineties: Problems, Themes and Meanings in Estonian Art on 1990s*. Eds. Sirje Helme and Johannes Saar. Tallinn: Center for Contemporary Arts, Estonia, 2001. 132-149. Print.

—. "Jagatud autorsus: kunstniku hajumine elektroonilistel väljadel [Shared authorship: dispersal of the artist in electronic fields]." *Kunstiteaduslikke Uurimusi* 16.3 (2007). 46-84. Print.

Kesküla, Kalev. "Sonetimasina mürinal." *Eesti Ekspress* 13 Jan. 2000. Print.

Kirby, Alan. *Digimodernism. How New Technologies Dismantle the Postmodern and Reconfigure Our Culture.* New York; London: Continuum, 2009. Print.

—. "Another Interview I Gave (Long, but Good I Hope)." 11 Aug. 2009. *Digimodernism*. Web. 6 Apr. 2010. http://digimodernism.blogspot.com/2009/08/another-interview-i-gave-long-but-good.html.

—. "Successor States to an Empire in Free Fall." *Times Higher Education* 27 May 2010. Web. 20 Dec. 2011. http://www.timeshighereducation.co.uk/story.asp?storycode=411731.

Kivisildnik. *kivisildnik oli siin*. Web. 11 Apr. 2010. http://kivisildnik.blogspot.com/.

Koskimaa, Raine. *Digital Literature. From Text to Hypertext and Beyond.* Jyväskylä: Jyväskylän yliopisto, 2000. Web. 6 Apr. 2010. http://www.cc.jyu.fi/~koskimaa/thesis/thesis.shtml.

Kris, Ernst, and Otto Kurz. *Legend, Myth, and Magic in the Image of the Artist: A Historical Experiment*. 1934. New Haven; London: Yale University Press, 1979. Print.

Krull, Hasso. *Trepp*. 1996. Web. 10 Apr. 2010. http://www.eki.ee/kodud/krull/.

Morrison, Aimeée [Aimée]. "Blogs and Blogging: Text and Practice." *A Companion to Digital Literary Studies*. Eds. Ray Siemens and Susan Schreibman. Oxford: Blackwell, 2008. Web. 25 May 2011. http://www.digitalhumanities.org/companionDLS.

Naremore, James. "Authorship." *A Companion to Film Theory*. Eds. Toby Miller and Robert Stam. 1999. Malden, MA; Oxford: Blackwell, 2004. 9-24. Print.

Pilv, Aare. *Pilve koht*. Web. 11 Apr. 2010. http://aaree.blogspot.com/.

Rettberg, Jill Walker. *Blogging*. Cambridge; Malden, MA: Polity, 2008. Print.

Rosenberg, Jim. "Questions about the Second Move." *Cybertext Yearbook* 2002-2003. Eds. Markku Eskelinen, Raine Koskimaa, Loss Pequeño Glazier and John Cayley. Jyväskylä: University of Jyväskylä, 2003. 83-87. Print.

Ruitlane, Olavi. *Olavi aeg*. Web. 11 Apr. 2010. http://ruitlane.blogspot.com/.

Sarapik, Virve. "Estonian Art of the 1990s and the Postsemiotic Turn." *Ülbed üheksakümnendad. Probleemid, teemad ja tähendused 1990. aastate eesti kunstis / Nosy Nineties: Problems, Themes and Meanings in Estonian Art on 1990s*. Eds. Sirje Helme and Johannes Saar. Tallinn: Center for Contemporary Arts, Estonia, 2001. 279-313. Print.

—. “Semi-Closed Openness: Transformations of Estonian Art and ‘Meta-Art’ in the 1990s.” *The End of Autonomy? Studies in Estonian Culture*. Studia Fenno-Ugrica Groningana 5. Ed. Cornelius Hasselblatt. Maastricht: Shaker Publishing, 2008. 69-84. Print.

Sarris, Andrew. “Notes on the Auteur Theory in 1962.” *Film Theory and Criticism: Introductory Readings*. 2nd ed. Eds. Gerald Mast and Marshall Cohen. 1962. Oxford: Oxford University Press, 1979. 650-665. Print.

Stam, Robert. *Film Theory: An Introduction*. Malden, MA; Oxford: Blackwell, 2000. Print.

Tamme, Tambet. *The Weather Station Never Lies*. 2002. Web. 11 Apr. 2010. http://thepression.tfd.ee/ulme/index.html.

Väljataga, Märt. *Sada tuhat miljardit millenniumisonetti*. [Tallinn]: Vagabund, 2000. Print.

Õnnepalu, Tõnu. Esna aeg. Web. 27 July 2012. http://orrisaar.blogspot.com/.

Piret Viires, PhD, is Professor of Estonian Literature at Tallinn University and Senior Reseacher at the research group of Cultural and Literary Theory of the Estonian Literary Museum. Her major areas of research are contemporary Estonian literature, postmodernism and post-postmodernism, digital literature. She is a member of AILC/ICLA Research Committee on Comparative Literature in the Digital Age.

Virve Sarapik, PhD, is a Senior Researcher at the Estonian Literary Museum and Professor in Art Theory at the Estonian Academy of Arts. Her principal research interests include semiotics of culture, visual semiotics, and the relations between pictorial and verbal representation. She is also editor-in-chief of the academic journal *Kunstiteaduslikke Uurimusi* (Studies on Art and Architecture).

Digital Memory, Risks and Common Sense: Dilemmas in the Context of National Libraries

Janne Andresoo

Director General
National Library of Estonia
Tõnismägi 2, 15189 Tallinn, Estonia
janne.andresoo@nlib.ee

Mihkel Volt

Head of the Research and Development Centre
National Library of Estonia
Tõnismägi 2, 15189 Tallinn, Estonia
mihkel.volt@nlib.ee

1. Introduction

In any country the national library collects, preserves and provides access to documents published in this country – it is an institution safeguarding texts that are important to the society. For national libraries, the tasks of collecting/preserving and providing access are like two sides of the same coin: one cannot exist without the other. Collecting and preserving are activities that build the potential availability, ensuring sustainable access to texts. However, as the practice of national libraries increasingly involves digital texts, it is time to view the possible changes in the working principles of national libraries that make cultural heritage available. Have these principles undergone any changes in the present 'digital era' and 'risk society'? What has been changed and for what reason? Which organisations experience similar changes?

2. The risk society and digital memory

According to Ulrich Beck, who wrote a book titled *Risk Society: Towards a New Modernity*, first published in German in 1986, each society faces global technological dangers. He writes that our modern society no longer follows any birth rights but enables people to establish their position through their own choices and achievements. In this modern society we come across a new kind of endangered status which seems impossible to avoid (Beck 1-20). Considering the relations between digital technology and the society, the following question may arise: can we speak about the endangered status of cultural heritage in relation to digital preservation and the provision of access? When thinking about preserving information, to what extent can we trust the technology that has developed and spread globally over the last 25 years? In this context the endangered status can occur both at personal and society level. For example, if all documents related to an individual are digital and the corresponding personal information is recorded only in digital databases, there is danger that a single technological error could delete all information proving this person's existence. Institutions that preserve national memory are facing a similar danger.

Today – in the 'digital era' – we mainly speak about 'digital memory'; but let us cast a glance also at the history of memories. Umberto Eco speaks about three types of memory:

> The first one is organic, which is the memory made of flesh and blood and administrated by our brain. The second is mineral, and in this sense mankind has known two kinds of mineral memory: millennia ago, this was the memory represented by clay tablets and obelisks, . . . on which people carved their texts. However, this second type is also the electronic memory of today's computers, based upon silicon. We have also known another kind of memory, the vegetal one, the one represented by the first papyruses, . . . and then on books, made of paper.

When transferring these thoughts of Umberto Eco to the context of national memory, we can say that in modern society national memory exists in a number of forms: oral, written, and audiovisual records. Each form of national memory – the so-called technology – and the attitude of society towards it have their own specific features. For example, the technology of oral tradition is the communication of information from one generation to the next. Although less than 100% of the folk songs that grandparents sing reach their grandchildren, such loss of information in the context of oral tradition has been tolerated as unavoidable. It is a matter of discussion as to what disappears more easily – is it a folk song on a less important topic or rather a song which is more difficult to remember. In any case it involves an obvious selection, either conscious or not. In the written tradition we know different information carriers (for example parchment, paper, digital form), as 'extensions' of human memory; different circulation ranges (manuscript form against published form); as well as collection and preservation principles (selective against exhaustive principles). The same categories can be used to classify audiovisual recordings.

In order to collect and preserve non-digital documents (and non-digital audiovisual recordings), archives and libraries have followed both the selective and the exhaustive principle; the latter, which aims to build up a comprehensive collection, is particularly characteristic of national libraries. Society has always expected that national archives and national libraries preserve the collected documents and publications permanently. Society has built up trust towards these institutions as keepers of national memory – meaning that society would not tolerate the loss of information that it has preserved. Memory institutions can keep this trust only if they are able to ensure the preservation of and access to the documents and publications they collect. Texts in digital form (born digital and digitised) are a specific type of written tradition. How should a national library organise the collection and preservation of digital content in order to justify and maintain the trust placed in this institution? There are suitable technological solutions for exhaustively collecting and providing access to both print publications and digital texts, for example online publications. However, the solutions for long-term preservation are very expensive, which means that selection cannot be avoided. In order to collect and preserve digital publications, national libraries have to fix the criteria with which to establish their meaning for the society, and compile a corresponding priority list which has to be the basis for collecting and preserving.

The attitude of the society towards the preservation of digital information is similar to its attitude towards oral tradition which accepts certain loss of information. Paradoxically, we may ask whether the society still expects that national libraries are able to guarantee permanent preservation of all publications, including 100% of online publications?

For example, the French web archiving specialist Julien Masanès (2006) considers it utopian that in selecting and organising web material – considering the extent of the web environment – a small number of librarians could replace the co-called publishers' filter. In his opinion, however, the organisation of web resources within clearly set limits is expedient and possible.

3. Principle of economics and national libraries

The centuries-long practice of national libraries put in a nutshell would be the following: each national library has collected and permanently preserved the possibly exhaustive print production of the country managing this library. This means that the national library collection includes each publication printed in this particular country. The corresponding conceptual initiative came from François I, the King of France, in 1537: a desire to create an exhaustive library of all books published in the Kingdom of France. The means for creating such an exhaustive library has been the legal deposit copy – known in the practice of national libraries from the times of François I onwards. This means that a national library gets at least one copy of each book published in its country. Collecting publications by means of deposit copies is all-inclusive: all titles are collected and preserved, there is no content-based selection. Which categories are to be collected – books, serials, journals, newspapers, maps, printed music, etc. – is at the discretion of each country and is thus specified at the national level. Such an all-inclusive approach to the acquisition of publications could also be interpreted as a principle of economics: it avoids the necessity to select national publications according to their contents, to develop the relevant rules and regulations, to hire staff. Collecting all titles of publication types designated as deposit copy objects is a general practice of national libraries, followed universally. Speaking of born digital publications, all-inclusive acquisition involves only off-line publications, e.g. CDs and DVDs. The practice of national libraries regarding the collecting and preserving of materials and making them accessible – based on the legal deposit system – has undergone a substantial change due to the appearance of born digital online publications, and also due to the digitisation of printed matter and other publications on analogue carriers to ensure better access to information and for updating it. Researcher Adrienne Muir from Loughborough University (2001) analyses the nature of the deposit copy in the digital environment and refers to Jan T. Jasion and his four criteria which the deposit copy should meet: 1) exhaustiveness, 2) ability to be preserved, 3) disclosure (mostly designating registration in the national bibliography), and 4) access. Exhaustiveness in the present context means the coverage of all information carriers as deposit copies regardless of their format or properties. Although there have been attempts to apply all Jasion's criteria to the deposit copy in the digital environment, there are still several legal, technical, economic and administrative questions which remain outstanding (Muir).

This is also an area where national libraries have been following the principle of economics, which means selecting the material to be collected and preserved on the basis of formal qualities (i.e. technical details) and/or contents as the technology used for digital preservation is expensive. It can thus be maintained that today national libraries follow two parallel principles for preserving publications of their home country, one aiming at maximum exhaustiveness, and the other being selective. Both these two acquisition principles and the activities related to digital content constitute a cultural change in national libraries. The selection practices in national libraries for collecting/preserving are somewhat similar to the appraisal of archival records in public archives. This similarity raises the question of whether national libraries should formulate internationally accepted selection principles for acquisition. Of what assistance to them could the assessment principles of archival records used in public archives be?

4. How long is 'long-term'?

As mentioned above, national libraries preserve born digital publications and digitise printed matter and other documents on analogue carriers in order to make them accessible over the web and thus give this information wider availability. However, the digital curation of documents is expensive and entails certain unpredictability: despite all efforts we do not know how many of these e-documents survive, for example, another 20 years or more. Yet we do know that paper produced for writing/printing since the industrial revolution of the 19th century is prone to become brittle due to its chemical properties. National libraries have several means to solve this problem, but their use to ensure the long-time preservation of information – part of national memory – is expensive. What are the choices for national libraries, considering that there is not enough money for everything that should be done to preserve national memory and make it accessible?

One factor influencing the decisions made by national libraries is the dominating preference for digital technology in relevant public discourses. In addition to the growing expense of digital content preservation, another problem is the preservation risks – imperfect technology and the ephemeral nature of digital preservation. What do we conceive as the 'long-term preservation of digital material'? Does the corresponding expert discourse define 'long-term' as a time frame of about 15-20 years? Yet, when speaking about the long-term preservation of paper materials, the same time frame is defined as a period of 75-199 years, and in the case of very slow decomposition, over 200 years. Digital technology is an excellent method for disseminating information, yet, regrettably, not the most reliable means for long-term preservation of information – at least not today. Why do we tend to idealise digital technology and eliminate any possible criticism of it? Memory institutions presently use combined solutions for preserving information in order to manage risks and ensure the provision of sustainable services; should they also contribute to protecting those technological solutions by providing explanations in public discourses? The question is how to ensure sustained and sufficient financing for the long-term preservation of the national memory?

5. Dilemmas of risk management

In the 20th century, libraries started to use different technologies to preserve printed publications, and to implement combined technological solutions for risk management in preservation and access. Libraries started to neutralise acid paper and to conserve publications on acid paper that had been used for printing since the 1850s. The national libraries of large and prosperous West European and North American countries have used the mass deacidification method for conserving printed matter on acid paper. Along with improving the original carriers, national libraries have also practiced the renewal of information. In order to reproduce publications to preserve information and make it accessible, publications have been microfilmed; for significant improvement of accessibility, the best solution has been digitisation. A national library thus preserves the information containing in national publications in 2 or 3 parallel forms: as print publications, microfilms and/or in digital form. Although microfilming has been considered the most reliable reproduction method for preservation, there are several practical considerations that question the feasibility of investing in this technology. Libraries have been facing the problem of unusable microfilms, the reasons for which are different – unsatisfactory quality of film developing, the incompatibility of older standards and standards regulating digitisation, or other reasons. This has lead to a new wave of reproduction. If something has happened to the copy – a microfilm or digital file – it is not easier today than in earlier years to make a new copy of a print publication the paper of which has not been improved by treatment. In the course of drafting "The strategic development plan of the National Library of Estonia 2009-2013" we recognised the necessity to foster both the long-term preservation of print materials and the preparation of digital copies as two reciprocally supplemental/supportive methods. In our opinion, the importance of digital copies lies in the fact that making the document conveniently accessible is the best way of protecting the original on paper. Digital technology enables the multiplication of information on paper with fewer resources than required by microfilming; digital copies are easier to transport to a duplicating preservation space in order to manage risks. However, the development of technologies also entails certain disadvantages – reproducing information from print materials by microfilming or digitisation enables us to save physical space, but we lose in the resistance to failures when ensuring stable preservation conditions, and in the resistance to *force majeure*. There is a better chance to preserve printed publications than microfilms and digital copies, e.g. in conditions lacking uninterrupted electricity supply and the preservation climate dependent on it. Comparison between digital forms and paper-based forms indicates that the identified deterioration of digital copies develops more quickly (Reed 119).

On the other hand, the advantage of digital copies over microfilm is the possibility to use technical equipment for convenient monitoring of the preservation of information – the actual readability of the text. Another advantage of digital copies is the possibility to use OCR and to create search options based on it. In addition, digital copies enable the retention and provision of easier access to the colours of the text and paper. True, we can also make colour microfilms but they are more expensive than black and white. Considering these factors, we have to admit that digital technology, despite certain disadvantages, is more functional and entails more options than microfilming.

6. What is the object of digital preservation?

In collecting and archiving online publications, the National Library of Estonia takes into consideration the selection principles for archiving online publications worked out under the Web Cultural Heritage project of the European Commission's Culture 2000 programme. The selection is governed by the general principle that the material should contain important information from the viewpoint of cultural heritage. Another factor to be taken into consideration is the specific character of natively digital material. Upon determining selection criteria for the collection and preservation of digital material, priority should be given to publications that lack a parallel print version. The majority of online newspapers and magazines are also published as print versions, but there are titles that exist only online. Studying the content of the Estonian web, for a particular group of online publications, it appears that all major newspapers with large circulation, both national and local, are represented with online versions. The archiving of online daily newspapers for long-term preservation is still a topic for discussion, the people involved with the issue try to find answers to the main questions: what would be the most expedient way to organise the archiving, and who should be responsible for it? Another question is what part of the digital information do we consider important, and how significant is the format of text material? Considering how simple it is to change digital information and to create different versions, it is essential to determine what the object of digital preservation is.

One of the problems related to digital material is the possibility of adequate preservation. The technology used to transfer digital information is becoming more and more sophisticated. Five years ago online newspapers consisted of static HTML files; today they are compiled dynamically. The advertising banner on the newspaper's margin constantly changes. A search in the newspaper's archive would produce an old article with today's advert on the screen. It is impossible to restore the screen image of the day when the article was published.

7. What kind of digital memory do we preserve?

In recent years the collection of pre-print files from Estonian publishers has been a positive experience for the National Library of Estonia. We have reached an agreement with a number of publishers that, in addition to the print edition, they also supply us with the original file sent to the printing house. For the library, such cooperation enables notable savings in resources – the immediate archiving of the digital version eliminates the necessity for subsequent digitisation or microfilming of this publication in order to preserve it.

In 2004 the National Library of Estonia started to create a digital archive for long-term preservation purposes. This was possible due to the library's participation in the reUSE project under the EU's E-Content framework programme. The main goal of the project was to create a digital archive environment and collect print files from publishers. As a result, the DIGAR (repository) archive was created using Fedora 2.0 software to collect, manage and preserve in the long-term electronic publications (digar.nlib.ee).

Under the national principles of bibliographic registration of Estonian publications, the following national publications are archived in the DIGAR digital archive:

- online publications published on the Internet, including books, newspapers, journals and magazines, serials, maps and printed music documents;
- digital copies of electronic documents published on physical carriers (e.g. floppies, CD-ROMs, etc.);
- digitised copies of analogue carriers, except audiovisual documents and pre-print files of print publications.

The digital archive collects Estonian online publications pursuant to the Legal Deposit Copy Act, acquiring publications through web monitoring or receiving them from the publishers; electronic publications on physical carriers are archived during processing. All electronic publications that have been submitted to the library as legal deposit copies are also archived retrospectively ("Sundeksemplari seadus", 1997). Each title is archived as one digital copy in the following preference:

- pre-print file;
- online publication;
- electronic publication on physical carrier;
- digitised copy of analogue carrier.

8. Conclusion: A new solution – provocative or reasonable?

During the drafting process of "The strategic development plan of the National Library of Estonia 2009-2013" we discussed the options of combining different technologies. One non-canonical idea was to create synergy by uniting the technological experience of the era preceding the industrial revolution with the experience of the digital era. Research has supplied us with knowledge that the deterioration of publications on cotton paper is slow or extremely slow (Aru).

Would it be possible to use print-on-demand technology for printing digital text on cotton paper? Would it be possible to provide a combined service of digitisation-on-demand and print-on-demand?

A hint from the past helped us find the answer to these questions. At the beginning of the 1920s the Estonian Ministry of Internal Affairs submitted a proposal for the draft Printing Act to print, on cotton paper, 8 deposit copies of each publication issued in Estonia in order to ensure their better preservation ("Trükiseadus I lugemisel"). The shorthand records of the session of the *Riigikogu* (Estonian Parliament) where the draft Print Act was discussed reveals that the said provision was not included in the draft Act on the proposal of Aleksander Veiler, a Member of Parliament and owner of a publishing and printing company at that time ("Trükiseadus II lugemisel").

The reason was obvious – in the 1920s it would have been difficult to execute this task both technically and organisationally. The digital era, however, has made it possible. This idea – seemingly rather provocative against the background of current practices – serves to draw attention to the following issue: if the tasks of national libraries to collect/preserve and provide access can be viewed as two sides of a coin, these two sides appear to contradict each other in the digital era and risk society. Providing access to digital content undoubtedly improves the possibilities to disseminate information more quickly, more easily and more widely. On the other hand, considering the issues raised in this article, everything pertaining to the collection/preservation of digital content drives us to ask the following:

Should we focus on the use of only one technological solution – digitisation – for the collection/preservation of information? Or should we reduce the risks of cultural heritage preservation by using more than one technological solution?

References

Aru, Krista, Kurmo Konsa, and Mari Siiner. *Rahvusteaviku säilivuse tagamine aastatel 2001-2004: projekti THULE andmebaasi 'Kultuuriväärtuste seisund Eesti suuremates raamatukogudes' analüüs ja rahvusteaviku kui terviku säilitamise strateegia väljatöötamine.* Tallinn; Tartu: Eesti Rahvusraamatukogu, 2002. Print.

Beck, Ulrich. *Risk Society: Towards a New Modernity*. London: Sage Publications, 1998. Print.

Eco, Umberto. "Vegetal and Mineral Memory: The Future of Books." *Al-Ahram Weekly* 20-26 Nov. 2003. Web. 20 Mar. 2012. http://weekly.ahram.org.eg/2003/665/bo3.htm.

Jasion, Jan T. *The International Guide to Legal Deposit*. Aldershot: Ashgate, 1991. Print.

Masanès, Julien. "Web Archiving: Issues and Methods." *Web Archiving*. Ed. Julien Masanès. Berlin; Heidelberg; New York: Springer, 2006. 1-46. Print.

Muir, Adrienne. "Legal Deposit and Preservation of Digital Publications: A Review of Research and Development Activity." *Journal of Documentation* 57. 5 (2001): 652-682. Print.

Reed, Barbara. "Challenges of Managing the Digitally Born Artefact." *Preservation Management for Libraries, Archives and Museums*. Eds. Gary E. Gorman, Sydney J. Shep. London: Facet Publishing, 2006. 117-132. Print.

The Strategic Development Plan of the National Library of Estonia 2009-2013. Web. 8 Mar. 2010.

"Sundeksemplari seadus." *Riigi Teataja I* 16 (1997): 495-499. Print.

"Trükiseadus I lugemisel." *I Riigikogu: IX istungjärk: protokollid nr. 185-221*. Tallinn, 1923. Vrg. 452-456. Print.

"Trükiseadus II lugemisel." *I Riigikogu: IX istungjärk: protokollid nr. 185-221*. Tallinn, 1923. Vrg. 1747-1754. Print.

Janne Andresoo, Director General of National Library of Estonia since 2008, started at this library in 1994 as bibliographic activities coordinator, later worked as Head of Research and Development Department. In 1987-1994 worked at Tallinn University as senior assistant and lecturer, teaching librarianship courses. In 1996 received MA in library science from Tallinn University. During doctoral studies in 2000-2004 at Tallinn University, her research interests included the selection principles of national bibliography and digital material in terms of long-term preservation.

Mihkel Volt graduated from the University of Tartu in the specialty of Estonian philology in 1989. Since 2008 he is Head of Research and Development Centre at the National Library of Estonia. His main interests are: the history of memory institutions, the theories and practices of cultural heritage preservation.

Cybertextuality Meets Transtextuality

Markku Eskelinen

Editor
Cybertext Database, http://cybertext.hum.jyu.fi
PL 276
00531 Helsinki
Finland
markku.eskelinen@kolumbus.fi

1. Introduction

The multitude of cybertextual media positions (Aarseth 62-64) implies many changes and challenges for traditional theories of inter- and transtextuality developed in the context of print literature. This paper will focus on certain novel types of relations existing between (and within) digital and ergodic texts and theorises them from a cybertextual perspective.

The paper has five empirical points of departure. First, ever since the emergence of the Internet, texts are no longer necessarily materially separated from each other and readers can seamlessly move between texts and chunks of texts that are mutually linked. Such movements are an obvious part of our everyday digital practice, but this type of relation (or should we say connection) is not recognised by traditional theories of inter- and transtextuality.

Second, traditional theories of intertextuality are fundamentally theories of textual relations that can only be interpreted (and not acted upon in any other way). Competent readers are expected to recognise the co-presence of texts within each other, or the way one text is modelled after another, but beyond recogni tion and interpretation (in the broad sense of producing meaning, significance and jouissance) they can do nothing with the relations they have either found or fabricated. In contrast, ergodic literature (especially text generators and textual instruments) often allows its users to affect, manipulate, and sometimes even create these relations.

Third, as printed signifiers are permanent, any relation between them, be it co-presence, commentary, imitation or transformation, is also permanent (although interpretations of the relations may vary). That does not have to be the case in digital media, where we have a field of (potentially) dynamic inter- and transtextuality instead of the old static one. Print texts are also inescapably intransient, and thus incapable of letting the users witness the real-time processes of textual transformations and replacements – for example from pre-existing source texts to pre-existing target texts (and vice versa) as in John Cayley's *riverIsland*.

Fourth, as print texts are closed volumes they can only quote, allude to, comment, imitate and transform prior texts. However, dynamic textual machines connected to the continuously expanding and routinely updated resources of the Internet such as *The Impermanence Agent* can or could also do this to the texts published after their own publication date. To a lesser degree this is also true with every text that accepts significant configurative textual input from the reader such as *Eliza* (1966–). Based on these observations we seem to have two major fields of transtextuality: open and closed, the former actively oriented not only to the past but to future texts as well.

Finally, we should be more careful to specify at what level relationships (and comparisons) between texts take place. In addition to the level of scriptons, relationships may also take place between textons and between the kinds of behaviours the compared texts display.

2. The choice of theory

Of course not every theory of intertextuality is equally responsive to these shifts, expansions and empirically verifiable details. The main choice is therefore between broad and narrow (or general and detailed) definitions of this contested concept.

One of the classic formulations of Kristeva (36) states that a text is "a permutation of texts, an intertextuality in the space of a given text" and within it "several utterances, taken from other texts, intersect and neutralize one another". In Kristeva's reformulation of Bakhtin's pet terms of dialogue and ambivalence the dynamic word's horizontal axis (sender/addressee) and vertical axis (text/context) coincide within the textual space resulting in intertextuality: "any text is constructed as a mosaic of quotations; any text is the absorption and transformation of another" (66).

From this point of view intertextuality is a normal and inescapable condition of all literature and textuality.

However, one could also be tempted to take one step further to more specific aspects and ask if there are any differences in the processes of construction, quotation, absorption and transformation, and whether it matters if texts quote each other explicitly or hide the relation, to be found only by readers competent enough, or if they imitate each other as parodies and pastiches or through other generic and stylistic templates, or can or cannot be enjoyed and understood without knowing their implicit references and models, and so on. In short, how much does it matter if the textual mosaic is not permanent and invariable but dynamic and transient, varying and supplementing itself with or without the reader's ergodic effort.

Genette's theory of transtextuality (Genette, "Palimpsests", "Paratexts" and "The Architext") provides a detailed overview of several important relations instead of using only one broad umbrella term, intertextuality, for all of them. To Genette ("Palimpsests" 1) the subject of poetics is not the text considered in its singularity, but transtextuality defined as "all that sets the text in a relationship, whether obvious or concealed, with other texts."

Genette ("Palimpsests" 1) finds five types of transtextual relationships: intertextuality, architextuality, paratextuality, metatextuality and hypertextuality. Architextuality is a purely taxonomic relation consisting of "the entire set of general or transcendent categories – types of discourse, modes of enunciation, literary genres – from which emerges each singular text." ("Palimpsests" 1)

Paratexts are heterogeneous elements that lie on the threshold of the text (such as titles and prefaces) and help to direct and guide the reception of a text by its readers (Genette, "Palimpsests" 3). They carry a pragmatic dimension, bearing and conditioning the text's "impact upon readers", and are thus interesting to us precisely because changes in textual behaviour are bound to affect these pragmatic threshold texts as well.

Genette ("Paratexts" 4-5) divides paratexts into epitexts (private and public outside texts that help readers interpret the text) and peritexts (the features literally framing the text). Instructions and users' manuals are the obvious contribution of digital and ergodic texts to the already recognised variety of paratexts, and the more complicated strategies the user has to employ in order to traverse the text, the more meaningful they become. In addition, any further information provided by epitexts about possibly hidden or inaccessible dimensions of the textual machine may have a higher status or relevance to the user than a more typical epitext potentially guiding interpretations.

Metatextuality is "the relationship most often labeled "commentary"" (Genette, "Palimpsests" 4). There are at least three types of new challenges ergodic texts may present. Firstly, in addition to the usual interpretative skills, the critic needs ergodic skills to use the textual system in an appropriate and successful way, and should perhaps be able to both explain and evaluate his or her strategy. Secondly, the potentially continuous and never-ending variation and supplementation may well exceed any humanly possible attention span, making the critic's observations and conclusions necessarily provisional. Thirdly, in certain encounters with indeterminate cybertexts, commentaries may turn out to be commentaries on one's own singular experience of ephemeral constellations of signs never to be repeated or to be seen by any other user, as ergodic works may be very sensitive to the way they are used.

According to Genette's ("Palimpsests" 1) restricted definition, intertextuality is "a relationship of co-presence between two texts or among several texts" including the practices of quotation, allusion and plagiarism. Genette's original formulation of hypertextuality ("Palimpsests" 5) is surprisingly broad and unspecific: "any relationship uniting a text B (which I shall call the *hypertext*) to an earlier text A ([…] the *hypotext*), upon which it is grafted in a manner that is not commentary." Genette later revised this rather negative definition as follows: "a hypertext is a text that derives from another by a formal and/or thematic process of transformation" ("Essays in Aesthetics" 10).

To Genette ("Palimpsests" 2-3), the difference between inter- and hypertextuality is also one of scale: the former is more local and optional taking place in semantic-semiotic microstructures "at the level of a sentence, a fragment, or a short, generally poetic, text", while the latter takes place between works considered as structural wholes. In his study of hypertextuality Genette concentrates mainly on what he calls "the sunnier side of hypertextuality: that in which the shift from hypotext to hypertext is both massive (an entire work B deriving from an entire work A) and more or less officially stated" (Genette, "Palimpsests" 9). This leaves open the darker side with more hidden and less massive relations between hypotexts and hypertexts – in short the meso level between intertextual micro level and hypertextual macro level, within which the relationship between the text and its reader is less openly contractual and pertaining to a conscious and organised pragmatics.

The five types of transtextuality are most of all mutually connected and sometimes overlapping aspects of all textuality, and only secondarily potential categories of texts (Genette, "Palimpsests" 9). For our purposes they provide a detailed enough framework of relations and practices against and within which cybertextual relations and possible novelties can be compared and theorised. Given the limitations of space, in what follows I will limit my observations to changes in inter- and hypertextuality.

3. Examples

1. Usually inter- and hypertextual relations are not directly announced and made explicit in the text; it is generally left to the reader to discover the relevant relationships. In many ergodic works this situation is reversed. In *Regime Change* (Wardrip-Fruin et al.) the user is presented with two texts that are named (a newspaper article on Saddam's fall in Iraq and the well-known Warren Commission report). Only the newspaper article forms the invariable part of the process (and hence serves as a hypotext to be transformed) and with each new session it is accompanied with a different chunk (quote) from the Warren Commission report. The main task is not to recognise the connections between the two explicitly

named texts but to mix them in interesting ways. Of course, both texts are already part of their own intertextual and generic networks, but these relations are not foregrounded and utilised in any way in *Regime Change*. From the outset the most obvious connection between the two texts is thematic: political violence. The two texts to be mixed or played in interesting ways are also non-fictional, which adds more emphasis and incentive to thematic readings and interpretations, perhaps providing simpler criteria (and a slightly lower threshold) for success. In *Regime Change*, as in many other instrumental texts, two or several texts are made co-present in widely varying ways in a process that doesn't have any given termination points – either to an individual session with the text or to the degrees of mastery to be achieved during multiple sessions. This kind of inter- and hypertextuality is both configurative and dynamic; the results of playing obviously vary resulting in different hypertexts quoting the intertext and transforming the hypotext differently every time.

2. Some digital texts, such as *The Impermanence Agent* (Wardrip-Fruin et al.), open up another dimension not theorised in the context of printed literature. *The Impermanence Agent* can react to, appropriate and use texts that were not available or didn't exist at the point of its publication. One could say that this is exactly what machines generally do as they are designed to deal also with future objects and processes.

The point is not the orientation towards future texts as such, but the nature of this relationship, i.e. that the earlier text will be the active party quoting and incorporating the later ones for its own purposes and by its own idiosyncratic rules. This operational novelty is based on the specific type of textonic dynamics of *The Impermanence Agent*: its ability to supplement itself with outside texts.

At the beginning of a process that lasts approximately one week the user reads the original story of Nana. During the process this fixed text will be gradually replaced so that in the end very little of it will remain. The replacement is based on material (texts and images) drawn from the user's browsing, but the precise logic of this replacement works in ways that are likely to escape the user even if he or she is aware of the general operational logic of the *Agent*. Browsing is necessary for the user's progress and the text's gradual replacement, but to view this progress the user needs to employ a peripheral mode of attention (Wardrip-Fruin et al. "The Impermanence Agent. Project and Context" 9): nothing happens to the *Agent*'s text if the reader doesn't browse between readings. The rhythm between reading and browsing is also left to the user: there are no paratexts suggesting how to fine-tune one's peripheral mode of attention, i.e. how often to take a look at the results.

With this work, two relationships are of interest: not only the relationship between Wardrip-Fruin's original story and the texts the user browses with or without giving it much thought, but also the relationship between the user and the agent. The user is free to visit the sites he or she wants or needs to, but it is the agent that decides what is important in them and how they will affect and be woven into the fixed original text. Ironically then, it is the human user who does the lower level (semi-intellectual) work serving the *Agen*t's decisions, and not the other way around.

The end results (final texts) of *The Impermanence Agent* are personalised according to the individual preferences and practices (as interpreted by the *Agent*) of each user. One can imagine two opposing poles of readership: ignorant readers making no exceptions to their normal browsing routines while the *Agent* is on their case, and experimental readers changing their browsing habits in order to tie the *Agent* to specific textual fields, modes and genres, or even individual texts, thus maximising their intertextual and hypertextual power, which will still be compromised by the *Agent*'s preferences.

3. *The Golden Lion* (Cayley) has two non-anonymous (i.e. explicitly named) seed texts, *Essay on the Golden Lion* by Fazang (643-712) and Cayley's short original poem *Han-Shan in Indra's Net.* Through a mesostic process the letters of a poem change one by one to the words of the essay (within the limits set by certain procedures of collocation) in a split screen presentation. The letters of the poem are shown in bold face within the words of the essay so it is easy or easier to follow the process. Here we can witness an intermodal transformation process between poetry and essay on the one hand, and the dynamic co-presence of Cayley's seed text within Fazang's essay within *The Golden Lion*. *The Golden Lion* shows intertextuality as a dynamic process in a way that broadens the possibilities for intertextual co-presence in two ways. First, transformations take place at the level of the letters, words and syntagms (accepted collocations) that determine each other in programmable ways that are unnatural in natural languages. Second, *The Golden Lion* both intertextually quotes and hypertextually transforms Fazang's essay in ways that distort its original syntax (and meaning) to a varying degree at the level of textual presentation (scriptons) while it quotes it (the whole hypotext) in full only at the level of textons (hidden from the user who can of course try to consult the original text by other means). There are many other lessons to be learned from Cayley's ingenuity, but for our current purposes these three are enough: the distinction between textonic and scriptonic inter- and hypertextuality, the scale or operating level of transformations and quotes extended to include letters (still visibly able to be combined into words), and the full presence of the hypotext on the textonic level of its hypertext.

4. In John Cayley's *riverIsland* (2008-) fully visible source texts (hypotexts) change into fully visible target texts (hypertexts) through transliteral morphing. Here we have intertextuality and hypertextuality as fully visible two-way processes instead of one fixed and permanent state of co-presence, imitation and transformation. The two interchangeable ends of this process (every text in *riverIsland* can function both as a source text and a target text) consist of 16 horizontal texts (Cayley's poetic adaptations of 16 poems from the 20 quatrains of Wang Wei's *Wang River Sequence*) and 16 vertical texts (different translations of one of the quatrains). Horizontal and vertical texts operated by the reader morph to each other letter by letter choosing the quickest and least resistant way through the alphabet, which means that the longest transformative chain consists of 14 intermediary letters as separate steps in the process. If we take into account only the source and target texts we have 512 (2 x 16 x 16) different hypertextual processes the user can launch and observe.

Moreover, the phases of the process wherein the target text is not yet fully arrived and the source text not completely disappeared give us a series of perceptible stages with their intertextual specifications as letter by letter; in addition to which the co-presences between the two chosen texts change and it is only at the beginning and the end of the process that there is no dynamic co-presence between the two. *riverIsland* quotes its 32 individual inter- and hypotexts in full and, unlike *The Golden Lion*, does not conceal them from the user. Alternatively, (if we shift the focus from the letter by letter transformations to the initial and final stages of the process) we could claim that the transformations in *riverIsland* between vertical and horizontal texts are not hypertextual, as the supposed hypertexts are not imitations or transformations (in Genette's sense) of the supposed hypotexts.

4. Conclusion

Within traditional scriptonic transtextuality, relationships can be static or dynamic, transient or intransient, and either interpretative or ergodic. To conclude I will revisit inter- and hypertextuality in more detail as within these relationships the changes are more visible than within the other three types of transtextuality.

On many occasion hypertextuality and intertextuality are tightly interwoven: hypotexts are often transformed by quotes from other texts. That's why we discuss inter- and hypertextuality together in the three dimensions mentioned above (dynamics, transience, and ergodics).

Dynamic (scriptonic) inter- and hypertextuality. Static co-presence is the norm in the otherwise conflicting theories of intertextuality, while there are many almost unexplored ways for the dynamic co-presence inherent in digital media to occur. Dynamic intertextuality can take place in three relationships:

1) The text's relation to its intertextual network (all the texts it either quotes or to which it alludes). Dynamic texts can incorporate quotations and allusions from new intertexts. This is what happens in *The Impermanence Agent*, which can quote whatever new site the user may visit and in the process of replacement some older references may vanish.
2) The relationship between the text and its particular intertext. If the relationship is static, then the intertext is always present exactly the same way in the text. If it is dynamic, then the intertext's co-presence undergoes changes. This is what happens to the quotes from the Warren Commission report in *Regime Change.* In principle, dynamic intertextual co-presence between a text and its intertext could undergo both quantitative changes (the area co-habited by the intertext may expand to the point of erasing and taking over the whole text or diminish to the point of vanishing entirely, i.e. the intertext may be quoted or alluded to more or less frequently and extensively) and qualitative changes if the original quotes are replaced by other quotes from the same intertext.
3) The presence and status of a particular quote or an allusion within the text may change. Overt intertextuality (quotes) may turn into covert (allusions) and vice versa. In addition, the position of a quote or an allusion in the text's composition may change and be either determinate (if the quote or allusion's adjacent scriptons are always the same) or indeterminate (if they are not).

Dynamic hypertextuality manifests first of all in the number of hypo- and hypertexts. Traditionally (as the relationship is between "structural wholes") the ratio is 1:1, one hypotext for each individual hypertext.

Of course there's no upper limit to how many separate hypertexts could be and have been constructed from one hypotext. Here the point is that several hypertexts and hypotexts can exist within the same text. For example, *riverIsland* contains 32 hypotexts and 32 hypertexts and 512 possible processes between them. In *Book Unbound* there is only one hypotext, although the number of hypertexts generated from it is unlimited.

Transient inter- and hypertextuality. Here the novelty lies in the way inter- and hypertextual processes are presented to the user. Traditionally hypotexts and hypertexts usually exist as separate volumes (although not necessarily: it varies because a hypertext can well contain a short hypotext) and they are certainly theorised as if this kind of relationship was the norm. In other words we have one static hypotext (as a permanent beginning situation so to speak) and its one static hypertext (one final outcome). Contrary to that doxa, the copresences between the two chosen texts in *riverIsland* change morph by morph, existing only as transient parts of the transformation process. In *The Golden Lion* Fazang's hypotext changes less frantically, one screen after another, in a one-way process.

riverIsland's two-way processes present another presentational variety by turning hypotexts and hypertexts into reversible positions giving the 32 individual parts of the text a double status as both hypo- and hypertexts.

Ergodic inter- and hypertextuality. Ergodic inter- and hypertextuality can be either explorative or configurative. In the explorative case the text is directly connected (linked) to its entire intertext or hypotext, which can be explored. The connections can be determined by the links in the text or by the user's browsing activity alone (as in *The Impermanence Agent*). Explorative forms are not particularly interesting as it is easy to let the user have complete or limited access to the intertexts and hypotexts to be used simply by including them in the text in their entirety.

Configurative inter- and hypertextuality can be further divided into impersonal, variable, and personal forms. Impersonal inter- and hypertextuality present the same texts and processes to every user. Whichever process the user happens to trigger while playing with the texts and the audiovisual streams of *riverIsland* there are no personalised differences or outcomes in them. Variable relations take place in *Regime Change* where they are bound by the given intertext (the Warren Commission report) and the given hypotext (the newspaper article), which are the only available source texts (to which the user can add nothing). *Eliza* offers possibilities for unbounded personal relations, as the user's input is not limited by any pre-given text.

Following the logic of the cybertextual user functions one could also conceptualise textonic forms of inter- and hypertextuality exemplified by *The Impermanence Agent* and *Book Unbound,* but here we would have to go back to the textonic level of transtextual relations that were already briefly discussed above.

Finally, dynamic transtextual relationships are also related to and intertwined with the text's dynamic intratextuality. Inner variabilities of cybertexts have in troduced a dynamic field of intra- or autotextuality, as texts could in principle consist of several more or less autonomous parts and phases quoting, imitating and transforming each other and changing their relations to each other. Under such conditions everything that has so far been theorised to take place only between texts could also happen within texts.

References

Aarseth, Espen J. *Cybertext. Perspectives on Ergodic Literature.* Baltimore: Johns Hopkins University Press, 1997. Print.

Cayley, John. *An Essay on The Golden Lion. Han-Shan in Indra's Net. Indra's Net IV.* London: Wellsweep, 1994. Electronic.

Cayley, John. *Book Unbound.* London: Wellsweep, 1995. Electronic.

Cayley, John. "riverIsland QT." *Instruments and Playable Text. The Iowa Review Web* 9.2 (2008). Web. 24 Feb. 2010. http://research-intermedia.art.uiowa.edu/tirw/vol9n2/artworks/riverisland/riverislandQT.html.

Eskelinen, Markku. "Travels in Cybertextuality. The Challenge of Ergodic Literature and Ludology to Literary Theory." Doctoral dissertation thesis. Jyväskylä: University of Jyväskylä, 2009. Print.

Genette, Gérard. *The Architext: An Introduction*. Trans. Jane E. Lewin. Berkeley; Los Angeles: University of California Press, 1992 [1979]. Print.

—. *Essays in Aesthetics*. Lincoln; London: University of Nebraska Press, 2005. Print.

—. *Palimpsests: Literature in the Second Degree*. Trans. Channa Newman and Claude Doubinsky. Lincoln; London: University of Nebraska Press, 1997 [1982]. Print.

—. *Paratexts: Thresholds in Interpretation*. Trans. Jane E. Lewin. Cambridge; New York; Oakleigh: Cambridge University Press, 1997 [1987]. Print.

Kristeva, Julia. *Desire in Language: A Semiotic Approach to Literature and Art*. New York: Columbia University Press, 1980. Print.

Wardrip-Fruin, Noah, Adam Chapman, Brion Moss, and Duane Whitehurst. *The Impermanence Agent*. 1998-2002. Electronic.

Wardrip-Fruin, Noah, Adam Chapman, Brion Moss, and Duane Whitehurst. "The Impermanence Agent. Project and Context." *The Cybertext Yearbook 2001*. Eds. Markku Eskelinen. Raine Koskimaa. Jyväskylä: University of Jyväskylä, 2002. 13-58. Print.

Wardrip-Fruin, Noah, David Durand, Brion Moss, and Elaine Froehlich. *Regime Change*. 2004. Web. 8 Oct. 2008. http://collection.eliterature.org/1/works/wardrip-fruin_durand_moss_froehlich__regime_change.html

Weizenbaum, Joseph. *Eliza*. Computer program. 1966-.

Markku Eskelinen, PhD, is an independent scholar and experimental writer of ergodic prose and critical essays. His most recent publication in English is *Cybertext Poetics. The Critical Landscape of New Media Literary Theory* (Continuum, 2012). Eskelinen is one of the founding editors of *Game Studies* – the international journal of computer game research.

From the Gutenberg Galaxy to the Internet Galaxy. Digital Textuality and the Change of the Cultural Landscape

Raine Koskimaa

Professor of Contemporary Culture Studies
Department of Art and Culture Studies
P.O. Box 35 (JT)
FI-40014 University of Jyväskylä, Finland
raine.koskimaa@jyu.fi

In this article I will address ongoing cultural change through the prism of digital textuality. The invention of hypertext, initially by Vannevar Bush, and later further developed by Douglas Engelbart and Ted Nelson, gave rise to a wholly new conceptualisation of what we understand by 'text'. The emergence of the Internet and the World Wide Web, combined with the hypertext mark-up language took hypertext to a global level. As a consequence, the cultural logic based on forms of print text, the Gutenberg Galaxy of Marshall McLuhan, is gradually giving way to the new digital logic of what Manuel Castells has termed the Internet Galaxy (Castells 2001).

1. Vannevar Bush – the prophet of hypertext

It is now commonly agreed upon that the 'father of hypertext' was the Director of the US Office of Scientific Research and Development (created by President Roosevelt in 1941), Dr Vannevar Bush, who published his highly prophetic article "As We May Think" in *The Atlantic Monthly*, July 1945. The article contains a wide range of speculation about future technological advances, and only a couple of pages are directly connected to the issue of hypertext (a word that did not then exist). Bush's main point was: the body of knowledge, scientific and otherwise, is growing so fast that something has to be done if it is to be useful at all. What Bush saw perfectly clearly was that the main problem lies in finding and using the information necessary at any given moment. Bush already took up the topic that has become widely acknowledged only after Internet was popularised: "Mere compression, of course, is not enough; one needs not only to make and store a record but also be able to consult it" (40).

Indexing had been in use in book format and libraries for centuries, and Bush also saw indexing as necessary but not satisfying alone. His addition to information retrieval was the introduction of recordable traces, paths, or links to the database. The user of *Memex* (as Bush named his hypothetical system) would have been allowed to make connections between different kinds and types of data and name those connections; a group of similarly titled connections formed paths. He also would have been able to add his own notes to the database and those additions could naturally have been added to existing paths. Thus, returning to the database, the user could follow his previous associative path through the materials. Bush uses a lawyer as his example: "The lawyer has at his touch the associated opinions and decisions of his whole experience, and of the experience of friends and authorities" (46).

The possibility to link different documents, to gather links to named paths, to add new documents to the database, and finally, to return to the database and follow the previous paths, are the key issues in any hypertext environment today. The strong role of the book form is quite obvious here: "It is exactly as though the physical items had been gathered together from widely separated sources and bound together to form a new book" (Bush 45). This use of the book metaphor has stuck to hypertext practice and theory to the present day, as witnessed in the highly confusing use of 'page' in World Wide Web terminology.

2. Ted Nelson – the visionary of hypertext

Theodore 'Ted' Nelson was the man who coined such terms as hypertext and hypermedia. In his book *Literary Machines* (first published in 1981, several revised editions since) Nelson gives a thorough description of his grand idea: "A program intended to make possible a new unified electronic literature [...] a computer program intended to tie everything together and make it all available to everyone", known as the *Xanadu* (the name is borrowed from Coleridge's poem "Kubla Khan" – but Nelson also mentions the connection to the great never-finished palace in Orson Welles' *Citizen Kane*) (Nelson 1/30 and the chapter "Beyond names, companies and products").[1]

Nelson makes clear his indebtedness to Vannevar Bush: "I say Bush was right, and so this book describes a new electronic form of the memex, and offers it to the world" (1/5). *Xanadu* was a plan for an electronic form of *Memex* – a database with user-defined links, paths and notes – although it was also something much more. While *Memex* in principle was a local machine (although the records might have been transferrable), *Xanadu* was a global system ("a unified concept of interconnected ideas and data").

1 Ted Nelson had already noted this hypertextual quality in *Pale Fire*: "[...] I believe in 1969, I arranged permission from the publishers of Nabokov's *Pale Fire* – a brilliant poetic hypertext – to use it in the demo. But the IBM people rejected this as "too far out." Thus progress must wait for the halt and the lame to catch up." (Nelson 1/31)

Nelson defines hypertext as "non-sequential writing":

> Well, by "hypertext" I mean *non-sequential writing* – text that branches and allows choices to the reader, best read at an interactive screen. As popularly conceived, this is a series of text chunks connected by links which offer the reader different pathways. (0/2, emphasis in original)

He also states that "Hypertext can include sequential text, and is thus the most general form of writing", and adds in a footnote, "In one direction of generalization, it is also the most general form of language" (Nelson 0/3). The improvement of writing in hypertext means that authors can write more freely, more flexibly, hypertext writing bending better to accommodate the forms of the topic in question. In addition, readers may more freely follow their interests when reading hypertext.

Nelson's grand idea was to build a global network in which individual hypertext documents could be linked to each other when necessary – in his own words, to make "instantaneous electronic literature" possible. Networked computers were seen as a platform for the most general form of writing – hypertext – thus, he termed computers 'literary machines'. There is also the stress on the referential nature of literature, which comes very close to the poststructuralist text concepts of Jacques Derrida and, especially, Roland Barthes. What to Barthes was simply 'text' is for Nelson the *docuverse*. Docuverse is the grand total of all documents in the global network, "a single great universal text and data grid" (Nelson 2/53).

All in all, Nelson clearly follows in Bush's footsteps when it comes to the enlightening role of the docuverse:

> electronic publishing will mean lower-capital entry (and thus smaller publishers), constant revisability of all documents with linked version update, and finally *open hypertext publishing* – the growing-together of a great jungle of interconnections among symbiotic documents, under separate ownership, becoming inseparable from the greater whole. And it means building systems for a hypertext and hypermedia constituency whose apex will be the most sophisticated readership and usership civilization has yet seen. (Nelson "Preface")

> The system … may or may not work. But some system of this type will, and can bring a new Golden Age to the human mind. (Nelson 0/12)

> Imagine a new libertarian literature with alternative explanations so anyone can choose the pathway or approach that best suits him or her … imagine a rebirth of literacy (Nelson 1/4).

Comparing this utopian view to the reality of the Internet today gives us quite a lot to think about.

As powerful a tool as hypertext is, it does not cover all the features inherent in digital media. For a better grasp of this potential, I will introduce the concept of cybertextuality. As defined by Espen Aarseth, cybertextuality refers to programmed texts that involve computation and which behave in various dynamic ways. Cybertexts are machines, which need to be operated by a user. This operating may take the form of navigation, modification, or participation. These are exactly the main modes of action we are witnessing in contemporary digital culture.

3. Cybertexts and ergodic literature

Cybertextuality, as Aarseth defines it, is a perspective on all texts that pays close attention to the ways they function. Texts in cybertextual scrutiny are understood as machines, not in a metaphorical, but in a very concrete sense. Cybertextuality is by no means limited to digital texts only – print texts also have their own ways to function (Aarseth 1-22).

Another important thing to note about the concept of cybertextuality is that it is not only about 'literature', not even in the broadest sense of the word. Cybertextuality includes forms like computer games and MUDs. When dealing with literature, Aarseth introduces another new term, ergodic. Ergodic literature requires 'non-trivial effort' to traverse a text. Following the lines by eye-movement and turning the pages are classified as 'trivial' efforts, anything additional as 'non-trivial'. Thus, books like Raymond Federman's *Double or Nothing* (1972), in which text is occasionally printed to circle around the page and thus force the reader to turn the whole book a full 360 degrees, can be seen as an example of ergodic literature. Vladimir Nabokov's *Pale Fire* (1962), in which the reader has to jump between the poem "Pale Fire", the "Commentary", and the "Index", is a case of ergodic literature, etc.[1]

In describing ergodic literature we need the concepts of *textons* and *scriptons*. *Textons* are the 'building blocks' of a text, the deep structure. *Scriptons*, in turn, are the possible combinations of *textons*, the surface level as seen by the reader. With any given hypertext, all the individual *lexia* (nodes) together are the *textons*, all the combinations of them, as chosen by any individual reader, are the *scriptons*. Or, with print texts like Marc Saporta's *Composition No. 1* (1962) where the text is printed on loose pages meant to be shuffled and read in random order, each leaf is a *texton*, all the orders in which they happen to be read are *scriptons*.

To accommodate how the reader has to participate in the meaning production of any text, Aarseth has coined the concept of *user function*. Basically, what it comes down to is to seriously think about the meanings of the term 'interactivity'. As should be evident, at least since Roman Ingarden's writings, all literature is interactive – the literary work of art is not completed without the active participation of the reader. There are many ways in which the reader may participate in literary meaning construction, and Aarseth has divided these into four basic user functions. The functions and their relations to other related concepts are illustrated in the following diagram:

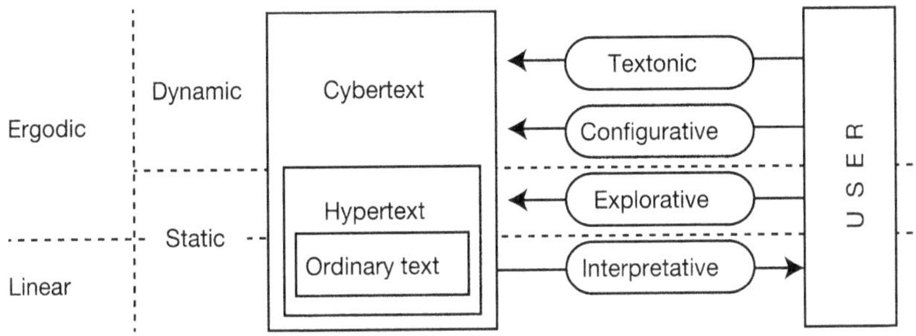

Figure 1. The user functions as defined by Aarseth (64).

Interpretation is the most basic user function: all texts require interpretation. *Explorative function* means choosing one's path through the textual materials available (or, in other terms, constructing scriptons from the available textons). *Configurative function* means the reforming of the textons or their relations (linking) in the personal document. *Textonic function* means the possibility to actively participate in the writing of the texts – by writing additional text, changing pre-existing text, or by deleting it; writing, it should be noted, may also include programming. In the popular usage of the term 'interactive' the explorative (or, selecting) function seems to prevail. In the previous discussions of hypertext literature the textonic function has usually been foregrounded – with hypertext literature "the reader becomes the author" has been a popular phrase. However, as the diagram above illustrates, the textonic function (which allows concrete co-authorship) does not belong at all to the domain of hypertexts! As Aarseth writes, "in a hypertext, the author can make sure that the user must go through a particular sequence to access a certain part...to suggest that the user is able to determine the shape of such a text is the same as to confuse the influence of a city's tourist guide with that of a city planner" (138-139).

Cybertext theory is most of all about the cybernetic production of signs and the unique dual materiality of that production. This uniqueness stems from arbitrary relations between two separate material levels (the storage medium and the interface medium) and leads to a heuristic typology of texts into which Aarseth is capable of situating every textual object from *I Ching* to MUDs based on how its medium functions, although independently of what that medium is. Cybertext theory is more interested in what a medium does than what it is and instead of more or less accurate ontological descriptions it gives us a continuum of media positions. This kind of approach could be called a functional theory of media.

Through examples from a set of digital texts, I will illustrate the characteristics of various phases of transitional digital culture. *These Waves of Girls* by Caitlin Fisher is a prime example of hypertext fiction. *The Impermanence Agent* by Noah Wardrip-Fruin et al. combines biographical writing with software agent technology monitoring the reader's online activity, creating collages of found online materials. The result reveals some of the fundamental features of writing (and reading) in a networked environment, especially various forms of citation and collaboration, both between human authors, and human and machine. *The Google Poem Generator* by Leevi Lehto employs a search engine for poetic composition. The generator highlights the utmost importance of search engines in negotiating the vast universe of online data.

4. *These Waves of Girls*

These Waves of Girls, by Caitlin Fisher won first prize in the 2001 fiction category awarded by the Electronic Literature Organization. In the category of electronic literature, Fisher's work can be further characterised with such labels as web fiction, hypertext fiction, and multimedial fiction. *These Waves* is published in the Internet and is available through Internet connection and a web browser. As such, it situates itself in the huge docuverse of the Internet – even though no links from the work reach outside its self-contained whole; through the web browser functionality it is always just one click away from other documents in the web. The work employs basic web site solutions, such as menu-based navigation and a page split into separate frames.

Even if it is a question of fiction, as with *These Waves*, since it uses the autobiographical form readers suppose that the narration is truthful *in the fictional world*. But the narrator of *These Waves* deliberately makes this assumption questionable. At one point she teases the audience by asking herself if things really happened the way she tells them.

> The desire to write is the desire to fool you, seduce you. Here I am – again – always getting the girl, saying the right thing or (toss this in for effect) something deliciously, winsomely wrong. Look over there – that's me, at four... (desire_to_write)
>
> I write, but it doesn't need to be my life, exactly. It lets me fill in the parts I forget. One name. One moment. A hand on my thigh that reminds me of all the other hands. Of yours. (hand_on_my_thigh.htm)

The narrator asks if it is believable that she really was that successful in her courting with other girls. Wouldn't it rather be that she tells *how she hopes things would have been*? Thus we as readers have to decide if we want to read the story as a realistic narration in which things have generally happened the way they are told, or if we treat the set of small stories as figments of the narrator's imagination, a narrator who is just making it all up as she goes. This is a classroom example of unreliable narration, as it forces us to ponder the reliability of text we read.

What is especially relevant for our discussion here is the way in which this whole thematic is buried deep in the hypertextual structure of the work. It is quite possible that the reader never faces the question of the reliability of the narrator, and in fact, it really requires quite a lot of patience to ever find these two nodes explicitly posing the question (quoted above, "desire_to_write" and "hand_on_my_thigh.htm"). If one goes through the labour of systemically charting the hypertextual structure of *These Waves*, one quickly sees that the network of nodes and links pretty well follow the basic principles of non-directional networks. Certain nodes are multiply linked to other nodes, with "Mr_Andersen", and "Vanessa", for example, both having more than ten links leading out from them, although the majority of the nodes have many fewer, but still several, links leading to and from them; and then there are some nodes with only one link leading to them. Both "the_desire_to_write", and "hand_on_my_thigh" belong to those almost isolated nodes. When reading *These Waves*, the reader frequently finds him or herself faced with "Vanessa", whether desired or not. However, the reader is required to stumble to the one link out of "Butterfly" in order to visit "desire_to_write" even once. Here we are facing a situation wholly different compared to reading traditional print fiction, in which pages and paragraphs are similarly available.

In print fiction, the underlying assumption is that a story is read in a linear manner from start to finish, and everything in between. In hypertext fiction this assumption does not hold. The whole idea of hypertext is that there is no linear text, but rather a set of interconnected text chunks. Thus, it is problematic if there really are such things as a beginning and end, although it is quite seldom the case that the reader really reads all of the text contained in the work (in many works it is actually impossible to read all of the text, as some text chunks may be made mutually exclusive: reading one makes another unavailable). This means that there may be readings in which this teasing about the reliability of the narrator does not figure at all. This, in its turn, takes us to the concept of dramatic irony, to a situation in which part of the audience is totally ignorant of something (the doubts the narrator casts on her own reliability) and reads the narration at face value, and another part of the audience does notice this modifier, simultaneously getting a richer understanding of the text and of the narrator, but also enjoying the extra pleasure of knowing that there are also ignorant fellow readers who never got the whole point.

In print fiction dramatic irony generally depends on sophisticated techniques as part of which foreign languages, complicate reasoning, references to other works of art, etc., divide the readership into those who get the joke and to those who do not. While these techniques, naturally, are also available to a hyperfiction author, he or she may additionally use the hypertextual strategy, hiding something in the hypertextual structure of the work. It may happen quite simply as here (a text node with only one link leading to it), so that superficial readers are left out of the revelation (although even then they can accidentally get to the crucial text fragment); or in a more complex manner a situation could arise in which, for example, a certain set of text fragments has to be read in a certain order to get to the desired result (as in Michael Joyce's *Afternoon*). Behind all this irony, however, *These Waves* is, finally, a very serious *Bildungsroman*.

5. *The Impermanence Agent* and critical technical practices

The Impermanence Agent by Noah Wardrip-Fruin et al. is a very interesting piece of digital literature that highlights several important aspects of cybertextual writing. Described very briefly, *The Agent* is a combination of a narrative text with illustrations and an agent-programme monitoring the WWW traffic of the machine on which the piece is run. The narrative is a biography, the memoirs of Wardrip-Fruin's grandmother, written down by Wardrip-Fruin and illustrated with pictures taken from the family album. This memoir is shown in a small window, which is supposed to be open on top of a web browser and possibly other running applications. The text proceeds automatically at a quiet pace, new 'pages' appearing after a set interval. While the programme is running and the text proceeding, the agent-programme continuously scans the WWW traffic of the computer. According to a certain procedure, the programme selects some of the web pages the reader/user has recently visited, cuts parts of those pages (fragments of either text or images) and pastes these fragments on the memoir text. Thus, the longer *The Agent* has run, the more of the original memoir is replaced by materials borrowed from the web pages visited by the reader. At some point, then, all of the original materials appear erased, a collage of borrowed web materials having taken their place.

We may observe here some quite obvious intentions. First, there is the issue of memories, how they gradually fade away, more recent impressions taking their place. In addition, we can see a juxtaposition of two representational logics, the narrative story giving way to 'sampling', or cutting and pasting, as a dominant mode of digital media communication. Further still, there are the traditional photographs that refer to a certain reality, serving to witness the once-existence of certain people and events (Roland Barthes has discussed this in detail in his *Camera Lucida*, 1981) contrasted with the pixelated digital images, often totally without a referent in the physical world. *The Impermanence Agent*, then, may not be the current world's *Remembrance of Things Past*, but rather the opposite, a demonstration of how the world gone by is quickly being replaced by new cultural formations.

There is another dimension to the work, however, about which its authors have written extensively in an article (Wardrip-Fruin et al.) in which they take up the notion of critical technical practices (CTP), a design principle advocated especially by a Swedish design movement. A central task of CTP is to make the users of a certain technology aware of the technology itself. *The Agent* employs the CTP approach in at least two ways. First, it makes the reader/user aware of his or her Web browsing routines. It is quite common that much of our daily web browsing is routine-driven, almost unconscious; we visit certain news sites, etc., without paying too much attention to them, not really registering what we are seeing, unless we encounter something specifically interesting. But when *The Agent* cuts fragments out of these pages, and pastes them in a new context, on top of its own original materials, we may pay attention to them in a new way – we may start pondering, for example, where is that fragment really from, from what page is it taken, to what topic it is really connected. Thus, it may make us more conscious of our own web browsing habits and so, in a sense, aware of our own unfolding 'biographies' and the materials by which our memories are being constantly replaced.

The very functioning of the agent-programme, on the other hand, is apt to show us that the monitoring of our web browsing is technically possible and, actually, quite simple. This is something most of us know, in principle, but it is very easy to forget in practice, or at least to comfort ourselves with the notion that nobody would care to expend the energy required to monitor us, out of all the millions of web users. Experiencing *The Agent* forces us to face the fact that, whatever we do in the web, we always leave traces of our actions in various servers and caches and it doesn't take more than a relatively simple agent-programme to monitor and filter that data. Here it is used for artistic purposes, visible only to ourselves, but it should help us to recognise the control potential inherent in networked media.

Furthermore, *The Agent* is also connected with the tradition of Dadaistic, Surrealistic and other avant-garde art procedures. This cybertext may be seen as a kind of automatic writing, a kind of semi-automatised 'exquisite corpse', a technological collage, a descendant of the earlier computer-based story and poem generators (such as the Racter by William Chamberlain and Thomas Etter), or even a variation of procedural text production along the lines of the Oulipian strategies. Finally, we may see it – after N. Katherine Hayles – as an instantiation of the post-human condition, literary production and meaning-making in a complex of human and technological agents. The final product is not the work of Wardrip-Fruin et al., nor of the user/reader, nor of the agent-programme, but of the combined efforts of all of them, with the addition of the unknown authors, designers and illustrators of the websites from which fragments have been cut, and even further, of all the technologies required to make the WWW tick. In Deleuzian terms, the author of this work should be seen as an *assemblage* of several actors, some of them human, some technological.

6. The Google Poem Generator

The Google Poem Generator by Finnish poet, translator, editor, publisher, etc., Leevi Lehto is, at first glance, like any poetry generator. It has a set of lyrical forms from which to choose (for example, a couplet), and then you can type in a "seed", a word or phrase. The generator then employs the Google search engine, and according to its own algorithm, constructs a poem in the requested form, based on the search results. Here are some examples:

> Stockard Channing puts on six different accents and sticks candles on your head
>
> You're forced to gild kumquats
> and bay leaves and string
> them on a wire
>
> ...beats anything I've come up with myself recently.
>
> posted by Sonny Jim at 4:33 PM on August 20, 2004

> of years of people who have promised
> 9,041. Time: 19:37. Zone: Heightened
>
> unworkable devices analyzed to expose
> Green and Ugly Trouble in Paradise
>
> over time, and mass. The staff
> Motion – An Ancient Mystery
>
> machine). This is a very short notice
> of an isolated mechanical device
>
> Motion is an internet-only radio show
> and the Perpetual Motion Roadshow

motion machines has long been
Heightened Reality. Meanwhile, on

motion machine). This is a very short
and clubs stage sports words art

of people who have promised free
logged in you would be able to see

Speeding Through the Night. Click
to achieve perpetual motion date back

of years of people who have promised
9,041. Time: 19:37. Zone: Heightened

posted by SteelyDuran at 6:26 AM on August 20, 2004

What is interesting here, as I see it, is the way in which Google Poem Generator harnessed the grand total of the WWW as its source material. The generator is not limited by the materials pre-loaded into it by the author, but rather the ever changing and expanding contents of the online communication channels bring it fresh materials each moment.

David Wall was the first to write about the WWW as a cyborg author, and I'd see the Google Poem Generator as an excellent instantiation of that idea. Even though the algorithm of the generator stays the same, it avoids the common problem of text generators, repetitiveness of forms and contents. They tend to be too predictable to be aesthetically pleasing. Predictability is a direct consequence of their predetermined functioning, in which everything is fixed in advance (even though the number of possible outcomes may be huge). In comparison, we could claim that the Google Poem Generator has an unconscious, or subconscious, level provided by the global, collective data stream of the Internet. True to the spirit of the cyborg author, the WWW itself takes part in the creative process. The interest here lies wholly somewhere other than the hypertextual or cybertextual structures and processes, it is the emergent meaning arising from the phenomenon of the 'noosphere' in Teilhard de Chardin's terms, or, "the Earth clothing itself with brain". Of course, the Google Poem Generator itself is rather a banal work, but its implications are huge. It winks to the direction of the creative potential of the emergent forms of social and semantic web.

7. Facing the digital world

Adalaide Morris has made an important claim: "what we do and see does not match the inscriptional or representational conventions through which we think" (3). That is, we are dealing with the new digital technology in our daily lives, especially in communications and media use, this being what we "do and see". On a practical level, then, new media technology is with us in a very fundamental sense. At the same time, however, our conceptual categories and theoretical ways to try and grasp the world, "the conventions through which we think", are borrowed from the era preceding the digital age. I have tried to show in this article how even the hyper and cybertext theories, despite their importance in paving the way for understanding the new cultural condition, are largely based on such Gutenberg era concepts as reading, writing, text and book.

These Waves of Girls exemplifies recent forms of creative practice, which are caught in the middle of a changing paradigm. *These Waves* effectively employs the Internet and hypertext technology, but is mainly based on older textual strategies (like *Bildungsroman*). *The Impermanence Agent*, based on a procedural strategy of agent technology, already to a large extent escapes the conceptualisations of hyper and cybertext theories. It is an important work of art for its attempt to try and grasp the ephemeral nature of the online world, but also in aesthetic terms, for how it combines elements from traditional literary texts and newer forms of creative production.

The Google Poem Generator bears the traditional poetic forms as a sort of remnant from the receding Gutenberg era, whereas its main thrust lies in the logic of a search engine. It exemplifies literary production on the premises of the new technologies we routinely use. The best way to try and get on terms with "what we do" in contemporary digital culture is to turn our attention to these sorts of programmed, software-based works, which reflect, in their digital form, the everyday experience of the digitised and mediatised world. Turning our attention to works that do not operate on the premises of 20th century (or earlier) text oriented theories is required before we can hope to remodel our conventions of thinking so that they will fit with our current experience in the Internet Galaxy. We do not need to abandon the traditional forms of literature and writing, but we have to recognise and acknowledge that they are no longer the privileged form of cultural expression. We should be open to phenomena shaping, and taking place in, online communications channels, in the fora of the social media and fields of digital play. Entering these new fields may be frightening, but it is certainly exciting. For those who are born into digital culture, the value of Gutenberg era practices is not self-evident. The great challenge, as I see it, lies in reshaping the curricula and teaching practices on all levels to better fit the cultural condition of the Internet Galaxy.

References

Aarseth, Espen J. *Cybertext. Perspectives on Ergodic Literature*. Baltimore: Johns Hopkins University Press, 1997. Print.

Barthes, Roland. *Camera Lucida: Reflections on Photography*. New York, NY: Hill and Wang, 1981.

Bush, Vannevar. "As We May Think." *The Atlantic Monthly* 176.1 (1945). Republished in *The New Media Reader*. Eds. Noah Wardrip-Fruin, Nick Montfort. Cambridge, MA; London: MIT Press, 2003. 37-47. Print.

Castells, Manuel. *The Internet Galaxy. Reflections on the Internet, Business, and Society*.Oxford: Oxford University Press, 2001. Print.

Hayles, N. Katherine. *How We Became Posthuman: Virtual Bodies in Cybernetics, Literature, and Informatics*. Chicago; London: University of Chicago Press, 1999. Print.

Morris, Adalaide. "New Media Poetics: As We May Think/How to Write". *New Media Poetics. Contexts, Technotexts, and Theories*. Eds. Adalaide Morris, Thomas Swiss. Cambridge, MA; London: MIT Press, 2006. 1-46. Print.

Nelson, Theodore Holm. *Literary Machines*. Sausalito, CA: Mindful Press, 1993 [1981]. Print.

Wardrip-Fruin, Noah et al. "The Impermanence Agent. Project and Context." *The Cybertext Yearbook 2001*. Eds. Markku Eskelinen, Raine Koskimaa. Jyväskylä: University of Jyväskylä, 2002. 13-58. Print.

—. "The Impermanence Agent. Project and Context". 2002. Web. 9. Apr. 2010. http://www.impermanenceagent.com/agent/essay2/.

Raine Koskimaa works as a Professor of Contemporary Culture Studies at the Department of Art and Culture Studies, University of Jyväskylä. In 2010-2011 he held the position of the Academy of Finland Senior Research Fellow. Koskimaa conducts research in the fields of digital textuality, programmable media, and game studies.

Between Technology and Teleology: Can the Digital Age Embrace the Analogue Experience of Culture?

Farouk Y. Seif

Professor Emeritus
Center for Creative Change
Antioch University Seattle
2326 Sixth Avenue, Seattle, WA 98121, USA
fseif@antioch.edu

This chapter introduces a new theoretical framework for cultural transformation in the digital age by juxtaposing the competing yet mutually reinforcing ideas of *technology* and *teleology*. The theoretical framework draws from philosophy, semiotics, and contemporary systems thinking, substantiated by traditional cultural practices. Based on this framework, and by reframing the challenge at hand, the paper calls for an eco-humanistic understanding of cultural transformation and suggests that: 1) digital technology is best viewed as a means to an end, but where the means and the end reciprocate purposefully in an integrative circularity; and 2) feeling comfortable with the paradoxical and tensional relationship between *techne* and *teleos* can bring forward a sustainable and desirable future that transcends virtual reality into an authentic and enhanced cultural reality.

1. Introduction: Cultures and the lure of technique

Cultures stand at a critical junction in history. Although the power of technology has been influencing cultures and civilisations since the beginning of time, our digital age brings unprecedented challenges that human beings have never encountered before. Technological innovation has taken place in almost all aspects of modern society, integrating technology into everyday life such that virtual systems of human interaction become interchangeable with physical systems. In effect, as many proclaim, the digital seems to have swallowed the real. This interchangeability between the virtual and the real seems to generate both challenges and opportunities for people to engage with their cultural practices.

Although technology has become pervasive, conditioning our entire cultural experience, digital technology has become the new addictive habit in our daily conduct. Modern society confronts the burden of both the gifts and curses of digital technology, where the lines between real relations and virtual relations become blurred, and interconnections between cultural objects and their respective signs become trivialised. Since the beginning of the 19th century, blind faith in technological progress and rationality has led to a widespread belief that technology alone can provide human beings with a sense of purpose. Such belief in technology has been coloured as fulfilment by a new different god. Ironically, a sense of purpose would hardly be an issue in assessing technological progress as long as the idea of progress remained a vague mixture of "shallow empiricism and hollow optimism" (Jaki 12).

There are several distinctive characteristics of our digital age that have far-reaching benefits but unintended cultural consequences. One of the most interesting characteristics is that the idea of culture is becoming less connected to geographical location. The relationship of cultural identity and heritage to physical place and regionalism has gradually weakened as mass digital communication extends beyond cultural boundaries. Another interesting feature is that the young now teach technique and, in turn, lead the old instead of the other way around[1] (Kanter). Consequently, the ideas of "tradition", "legacy", and "heritage" in this so-called "pseudo-social life"[2] have generally become negative terms, implying an archaic cultural worldview that ignores technological innovation and hinders progress. For instance, Internet economic tendencies appear to be antihistorical, promising a premium for new ideas that have been deliberately removed from cultural traditions. In fact, since the time of Aristotle, technique has been viewed as synonymous with epistemology; consequently, technology has become the "new science" that must be taught by the masters of technique (Dunne). According to many critics, this view of technology as science-to-be-taught, which is inherently an American phenomenon, is considered to be an indication of the trend to disregard history and overlook the wisdom of the ages. Ironically, this disregard for history is counterproductive to the reality and historical interrelationship between culture and technology.

In addition, the emphasis of societies on technological novelty and the value of change for its own sake, in conjunction with a high regard for rationality and individuality, appears to foster idiosyncratic ideals that ironically work against themselves, since novelty and individuality can become meaningful solely as

1 This observation was originally made by the anthropologist Margaret Mead, who asserts that America is the first culture where the young children of immigrants learned the new language first and taught the old. See Rosabeth Moss Kanter. *Evolve!: Succeeding in the Digital Culture of Tomorrow.*

2 Although the term "pseudo-social life" implies a disingenuous-social life, however in the context of this paper, its meaning is intended to convey the nature of virtual-social life. It emphasises the difference between building community *online* and building community *offline*.

parts of the social constructive discourse. Individuals in the digital culture seem to be superficially interested in what is *new* and what is *next,* with little attention to or regard for purposeful interactions or meaningful interpretations of cultures. Consequently, the world we see is the globalising world of the commodity, where social space is blanketed with ever-new layers of technological commodities.

Even before the term globalisation was popularised, Guy Debord, in his book *Society of the Spectacle,*[3] offered an important critique of the mass media and technological consumerism that contribute to alienation of social life, ethnic divide, cultural homogenisation, and dilution of cultural identity. Not only does the spectacle produce a pseudo-social life, but it also directs society's attention to the *lure of technique.* The spectacle seems to eliminate the boundaries between the real and the virtual by relaying on illusory magical techniques.

The heart of the matter seems to be not only the tenuous separation of technical knowledge from practical and purposeful knowledge, but also the tendency of our digital age to regard technical knowledge as superior to practical wisdom. It is not a surprise to observe that a Western conceptual model has driven much of our understanding of the digital age. And, as we shall see shortly, this model, which has its roots in Greek philosophy, has inexcusably overlooked the idea of intention and purpose. The real issue, then, is not just *how* digital tools can facilitate users-centred changes in cultural transformations. Rather, the issue is more critically related to the *purpose* of these digital tools and how to intentionally engage users with cultural heritage and transformation. Focusing on ways to use digital tools for users to engage with cultural heritage seems to hinge on the *intention* that orients both users and designers to facilitate cultural transformations.

The above observations raise several questions that are helpful in organising our discussion and seeking more sensibility and understanding: In what ways does the digital age endanger or sustain cultural heritage? How can the information and communication technology help to preserve the identity of cultures in our ever-changing and homogenising digital world? How do we capitalise on the power of virtual reality to maintain cultural memory? What fundamentally different ways of thinking and interacting with the digital world are there that might enable designers and users to not only sustain heritages but also transform cultures? What roles do digital communications play in moving beyond the pseudo-social life to authentic cultural practices? These questions dig deep into our consciousness and challenge our cultural practices. In order to make an attempt to answer these questions, we first need to consider the recent debate on the digital age.

3 Guy Debord's *Society of the Spectacle* was first published in French in 1967. There have been several translations into English. The version cited here was translated by Ken Knabb and has no date.

2. Assessing the debate on the digital age

Each pole of the debate seems to have its own rationales, to a degree that one can easily become persuaded to join one of the two poles of the debate. However, by assessing and juxtaposing the two contrasting and, in someway, contradicting arguments, I hope to highlight some aspects of each pole that assist in revealing what really matters in the digital age. The debate among proponents and opponents of digital technology and its role in the transformation of culture seems to lead to confusion and frustration – confusion about the means-end framework, and frustration with our inability to reach a sensible understanding. As stated earlier, the digital age has created the most challenging and paradoxical situation in human history. Opponents believe that we have become infatuated with digital technology as a means of communication, making it easy to overlook the significance of the analogue experience of reality, which is seamless, undifferentiated, and whole. For them, digitisation is becoming our addictive path to maximisation. But this is the antithesis of the nature of life, because "life tends to optimize rather than maximize" (Hawken 183). Even though the analogue experience of reality is not particularly efficient, it is undoubtedly an effective way to sustain and optimise cultural memory.

Granted that social networks provide unprecedented resources and support for professional development and emotional needs; however, they also foster new social predicaments and they lack authentic dialogical relationships. It has been suggested by many critics that with unlimited resources and far-reaching mass communications, digital technology has become the new untamed beast that causes an avalanche of wicked social and cultural problems. The digital age exemplified in information and communication technology has even empowered a few individuals to mobilise large populations to affect gigantic corporations and influence governments of nations. The World Wide Web has grown to be the new town-hall meeting for the global village, where conversations determine the fates of nations and corporations. Digital technology lends itself to "guerilla tactics" such as the rapid spread of unconfirmed reports and leaked confidential documents (Kanter). Equally critical, while the use of technological innovation offers unlimited opportunities for sharing values and human rights, it brings with it a moral dilemma – how to protect the underage and juvenile from inappropriate content and from Internet predators and parasites.

Moreover, the digital age demonstrated by the Internet can empower and connect people within a culture and across cultures, but it can also segregate and marginalise them. The Internet can certainly encourage communities to form and grow, but it can also provide a powerful medium to deny, attack, and destroy other communities. Interestingly, the community of the website is really a mere vague analogy for potential interaction rather than emotional reality. According to critics, "community" is sometimes a distorted analogy because many so-called communities are just excuses for making money: while charming people into viewing, such sites are "monetizing" their "eyeballs" (Kanter 18). And despite all efforts to close it, the gap in the digital divide has expanded in recent years. Certainly, these inequalities among world populations suggest serious ethnic divides. At the same time, the coupling of the ethnic divide and globalisation spawns one of the most paradoxical situations – homogenisation and alienation – modern society has encountered since the time of the Industrial Revolution.

The illusory magical techniques have affected our entire set of cultural practices – exemplified in global exchanges of information, freedom of expression, and the unprecedented sphere of choices. Almost anyone anywhere is able to participate and be included in the benefits of the global information society. Many opponents of digital technology believe that the "real" world seems to be replaced by images that make themselves the epitome of reality, where technology is viewed as a factitious god, making its own rules and aims for nothing but itself. Whenever the real world is distorted into mere images, these images become real beings, a kind of self-motivated figment for hypnotic behaviour. And since the spectacle's task is to use a range of mediations in order to show us reality that can no longer be directly grasped, it elevates the sense of sight to special preeminence where a real dialogue becomes almost impossible (Debord).

Debord's argument echoes my graduate students' complaints about online discussions. For nearly two decades, graduate students have favoured the analogue experience that often occurs in face-to-face conversation over online discussions. Although online learning has reached beyond the classroom experience through teleconferencing, chat rooms, and the like, creating virtual learning communities for lifelong learning, it has also opened the floodgate for trivialisation and what can be perceived as "hit-and-run" intellectual activities; consequently, it has often missing real scholarly acumen. In-person communication is particularly significant for engaging in dialogue. There is no such thing as real dialogue through online interactions. Dialogue is more than thinking together, it is imagining together in real time; thus, the so-called "online dialogue" is only an illusionary cultural space in a shallow virtual environment. Online discussion, as it now stands, is characterised by serious limitations and shallow communication. Making an insightful connection between communication and community, Rollo May writes:

> Communication presupposes community which, in turn, means a communion between the consciousness of the persons in the community. This is a meaningful interchange which is not dependent upon the individual's mere whim, but is a built-in

> aspect of the structure of human intercourse . . . Understanding is possible, specifically by *the structure of language*, and more generally by *the structure of human relationships*. (156)

In the broad sense, technology is a medium of communication, a system of representations, capable of not only describing cultural objects, but also making it possible to modify them; and when it modifies these objects, it transforms itself. And this unlimited process of transformation or mutability generates cultural phenomena that are difficult to place under traditional genres or categories (Pinto). As a matter of fact, this is the basic dynamic of Peirce's semiotics (Peirce) that can be perceived as a virtual-actual semiosis, which makes the impossible possible. Undoubtedly, technology has impacted the way we see and understand cultural objects and practices; and consequently, it may require rethinking the nature of community. Thus online communities have the potential to match, or even supersede, offline communities.

The debate among proponents and opponents of digital technology seems to go nowhere. The actual and the virtual, what is online and what is offline seem to converge through signs (Pinto) into diverse, converse, and optimised experience. As indicated above, although the nature of life favours optimisation over maximisation, effectiveness rather than efficiency, ironically digitisation continues to be our path to efficient maximisation. And because of our habit of convenience, it is neither impossible nor undesirable for our technologically advanced society to resurrect cultural practices of the "good old days". An equally valid counterargument is that while technology has a powerful impact on how we view our cultural practices, accepting changes triggered by technological innovation has never been easy.

Considering these habits, and according to many promoters of the digital age, the technology that many fear will homogenise or dilute cultures may, in fact, encourage the opposite by serving as a vehicle to give voice to local cultures and indigenous populations, preserving their identities and restoring their cultural memories. Technology, then, is neither a constraint nor a limitation, since digital services, networking, wireless connections, instant communication, and knowledge management tools are sufficient to satisfy human needs, according to digital age proponents. However, as one of these proponents, Alfredo M. Ronchi asserts:

> The challenge now is how to take advantage of these tools by channeling their use into creating true innovation. We must work out how to positively influence society by making use of new opportunities, and how to reverse the *digital divide* in order to create *digital opportunities*. (437)

The mass rush to digitisation comes with imaginative results but also with a high price. For example, Mark Poster argues that "the media transform place and space in such a way that what had been regarded as the locus of the everyday can no longer be distinguished as separate from its opposite" (Poster 211). Digitisation, while it is an advantageous technological achievement in speed, accuracy, and efficiency, seems to have impoverished the role of the analogue mode in simultaneously capturing whole and seamless cultural experiences. Unlike the digital mode, which represents the complexity of life by breaking information down into discrete bits or packets to be easily compacted, stored, and mass-produced, the analogue mode consists of direct representational or analogous systems of continuous, undifferentiated, whole experiences. An analogue mode represents a continuous flow and undifferentiated movement of human experience. So the nagging question remains, can the digital age embrace the analogue experience of culture?

The answer seems to depend on paying close attention to the notion of *teleology*. The process of reversing, or rather reframing, the challenge at hand and therefore creating digital opportunities depends on our ability to acquire a deep understanding of *sense of purpose*. We need to keep in mind the overarching notion of teleology – to develop a clear understanding of the appropriate distinctions as well as connections between means and ends, techne and teleos. More specifically and interestingly, we need to recognise that the realms of techne and teleos partly coincide.

3. The interrelationship between *techne* and *teleos*

The term *techne* denotes technical knowledge; it reached great significance in Aristotle's writings. Because of the intimate connection between techne and rationality, the product of techne is likely to be clearly definable and concrete things; consequently, techne becomes attached to the concept of *teleos* – purpose and intention. Joseph Dunne draws a close connection between Aristotle's two forms of knowledge, techne (technical knowledge) and *phronesis* (practical knowledge). And in doing so, Dunne coincidently makes helpful remarks regarding the long-held belief that techne and teleos are intimately linked. However, as we shall see, techne and teleos are interrelated in the Aristotelian sense in what seems to be a peculiar way.

Techne itself is not a useful thing; rather it is a generative cause of useful things, argues Joseph Dunne. In the Aristotelian sense, techne is a kind of habitual skill of the craftsperson through which he or she can produce and reproduce practical things. Undoubtedly, there is a fascination with this kind of activity simply because the craftsperson is likely to produce clearly definable and tangible products, where success and performance can be achieved and measured. Although this fascination with techne, the art of making, has dominated societies for centuries, it has reached its pinnacle in our digital age. Often good judgment about the outcomes of technological innovation and the awareness of intentionality seems to have escaped human sensibility. The gap between technical knowledge and intentionality demands a critical examination.

First, let us explore further the concept of intentionality. Intentionality is the structure of meaning that makes it possible for us not only to perceive the world but also to make meaning of it. Arabic philosophers such as Ibn Sina and Ibn Rushd introduced the specific concept of intentionality in early medieval times, conceiving it as a significant principal in the thought of the Middle Ages (*Internet Encyclopedia of Philosophy*). At that time intentionality meant epistemology – how we know reality. Intentionality was twofold: knowing particular existing things or objects, and knowing the relations of these objects and things to general concepts – that is, knowing by conceptualisation (May). Thus, intentionality is revealed through conceptualisation as long as conceptualisation embodies contemplative reflections during the process of making or doing. In other words, while the process of conceptualisation reveals intention, contemplative reflections through the process of implementation make the appropriate adjustment and rectification for navigating toward a clear intention. In a sense, there is a feedback loop between the conceptualisation process and the implementation process[4] in an evolving circularity of meaning making. This feedback loop between conceptualisation and implementation seems to have escaped the attention of most designers of digital technologies.

Intentionality is also associated with vitality, aliveness, and the intensity of the analogue experience in the process of meaning making. And since meaning making is one of the most significant aspects of intentionality (May), human beings are not merely rational animals, they are "semiotic animals" (Deely), capable of engaging in the process of meaning making. John Deely writes:

> A semiotic animal is not an animal merely capable of knowing that there are signs . . . The semiotic animal emerges only as *realization* of the capacity of the rational animal for "perceiving the relation of signification". (114)

I believe this capacity of semiotic animals to perceive the relation of signification is intimately linked to what has been called "cunning intelligence". Joseph Dunne tells us, "the Greeks prized very highly a talent for making out against the

4 Observing my graduate students struggle over the years with the two different processes of conceptualisation and implementation revealed the challenging paradoxical interrelationship between techne and teleos. Design students who understood and reflected mindfully through these processes developed clarity about the design intention and achieved desired outcomes.

odds of greater strength" (257). And the French scholars Marcel Detienne and Jean-Pierre Vernant name this talent as "cunning intelligence". They describe this talent as that which integrates

> ... flair, wisdom, forethought, subtlety of mind, deception, resourcefulness, vigilance, opportunities, various skills, and experience acquired over years ... [and] is applied to situations which are transient, shifting, disconcerting and ambiguous, situations which do not lend themselves to precise measurement, exact calculation or rigorous logic. (Detienne and Vernant 3-4)

The above assertion about cunning intelligence is essential for our understanding of the interrelationship between techne and teleos. Since this kind of intelligence, or rather design intelligence, deals with uncertainty and ambiguity, its use requires human beings to develop a great deal of perseverance. Interestingly, but not surprisingly, the value of "perseverance" and the notion of "design" are beautifully connected beyond any expectations. The Arabic word *tasmeame* means perseverance or design. Design through perseverance embodies both techne and teleos; it is the way we deal creatively with the paradox of technological needs and human values. To engage meaningfully in the design of technological innovation, one must develop a cunning intelligence and a capacity for perseverance in order to thrive despite all odds in the challenging situation at hand.

4. Teleological interpretation for a technological world

Techne, the root word of technology, meant "art" – *the art of making*. But technology has recently been blamed for nearly every human difficulty: moral challenges, homogenisation of culture, exploitation of nature, loneliness, ugliness, and many others. This view triggers conflict between technological needs and human values. However, the way to deal with this conflict is not to run away from technology. That is impossible. Neither is it reasonable to use technology as the solution for social problems, or treat technology itself as a mere problem to be solved. That creates more messes (Ackoff). "The way to resolve the conflict is to break down the barriers of dualistic thought that prevent a real understanding of what technology is – not an exploitation of nature, but a fusion of nature and the human spirit into a new kind of creation that transcends both" (Pirsig 291). In fact, and based on the Heideggarian phenomenology of technics, technology – the art of making – can reframe our knowledge and experience, opening the possibility of transparency between nature and culture (Heidegger) and thus avoiding the tenuous trap of dualism.

Even those who question how the digital age sustains cultures may have overlooked the idea that "tradition may be thought of as what is *underneath* us, continually exercising a kind of gravitational pull. Not that it necessarily makes for a deadweight; for the ground to which it connects us by a thousand invisible filaments, in our thought and in our being, can be a sustaining one, making possibilities substantial and actual" (Dunne 359-360). Digital technology may have been the single most robust force beneath this gravitational pull of cultural heritage and environmental sustainability, which Paul Hawken articulates in his book *Blessed Unrest, How the Largest Movement in the World Came into Being and Why No One Saw It Coming*." Hawken concludes:

> The Internet and other communication technologies have revolutionized what is possible for small groups to accomplish and are accordingly changing the loci of power. There have always been networks of powerful people, but until recently it has never been possible for the entire world to be connected. (6)

Certainly, we do have the powerful means of digital technology to mobilise social change, engage in cultural transformation, and sustain cultural heritage. What we often overlook, however, is the sense of purpose. More significant, what we need is to develop a deep understanding of the interrelationship between technology and teleology. Perhaps the widespread and commonplace use of digital technology, but with mindfulness and contemplative reflection throughout the conceptualisation and implementation processes, can reveal a clear intention that promotes a sustainable way of life celebrating environmental sensibility and cultural sensitivity. Transformation of cultures in the digital age is neither unavoidable nor inevitable. In fact, a culture that ignores experimentation with technological innovation is destined to decay over time. Indeed, innovative ideas contain the familiar past and the permanent. Innovation may be the emergence of something utterly new, but it is also the rekindling of something always true (Grudin). On the one hand, cultural innovation is essential for social systems because technological innovation functions as a negative feedback loop in social systems for adjustment and renewal. On the other hand, stability and consistency of cultural heritage, as a positive feedback loop, are fundamental if society is to prosper and develop (Hall; Mumford; Seif).

While social constructionists view change as essential for human improvement, they overlook the fact that stability and continuity are needed for change to be recognised in the first place. Seeking novelty as an expression of economic progress, an idea inherited from Social Darwinism, has led social constructionists to believe that technology offers solutions to social problems. Based on Gebser's work, *The Ever-Present Origin*, change does not surpass the past; the "new" presumes the "old" through integration. And this integration can permeate the intersection of nature and culture, and the diaphaneity of techne and teleos.

To initiate an action for a technological invention, one must be clear about its driving intention. Note, however, that intention conveys a tendency to move toward something (May); and therefore, intention is a journey, not a destination (Seif). In our digital age, the pressing matter is to develop teleological clarity about what technological innovation is for. The approach of "tyrannical technology" (Ronchi 142) imposes its own view on both designers and users, forcing them to operate according to the needs of digital techniques and tools, and more often than not, engaging them in meaningless tasks. But this tyrannical technology is not the cause of the complicated interface, as Alfredo M. Ronchi claims; rather, the cause is the lack of a clear understanding of the purpose these digital tools and techniques are serving. As I mentioned earlier, there is the lure of, and infatuation with, the idea of technique for its own sake, which encourages a predetermined technological action without awareness of the unfolding teleological process. As Jean Gebser reminds us that nothing exists for its own sake; "it exists for the sake of the whole" (541). Digital tools and techniques are not autonomous entities and methods serving nothing but themselves.

Perhaps the most critical, but unnoticed forces of the digital age are widespread individualism (Kanter) and obsession with rationalism. These hidden forces obliterate the potential for teleological interpretations of technology in transforming and maintaining cultures. As stated earlier, human beings are more than rational animals, and such being the case, John Deely asserts:

> In the semiotic animal, the modern experiment has passed its limits with the realization that rationality is by itself not yet reasonableness, "for reasonableness is the capacity to respond to the attraction exerted upon the self by the other," not the isolation of the self from the other. (116)

Only through the meaningful integration of human systems and more-than-human systems forming a semiotic web can humans, as semiotic animals, make the appropriate distinctions between wise use and reckless abuse, realising that unintended consequences may extend well beyond social and cultural interactions (Deely). This principle is at the heart of *semioethics*, which transcends the notion of "semiosphere" originated by Yuri M. Lotman into a co-evolutionary process of meaning making.

The self-evolutionary process that Erich Jantsch talks about is diametrically opposed to the tendency to use technological solutions to mitigate social and cultural problems, encouraging individualistic attitudes toward social change as autonomous human affairs (Seif). Transformation of culture in the digital age hinges on the intentional interaction between technological innovation and cultural practices. In other words, cultural transformation depends on our ability to recognise that the realm of techne and the realm of teleos are diaphanously coinciding. More specifically, engaging authentically in the process of technological innovation and the art of making (techne) brings forth a desirable cultural transformation, which in turn reveals purpose and intention (teleos). This intentional interaction with digital tools and techniques is precisely how we can discern between technology that endangers and technology that sustains cultural heritage.

5. Conclusion: Implications for cultural transformation

It is nearly impossible for our present societies to judge their own time in history. The persistent question about whether the digital age can embrace the analogue experience of culture remains unanswered. Perhaps the answer not only hinges on our perseverance through paradoxical tension but also depends on our sensibility to overcome the duality of techne and teleos. This sensibility has the potential to foster an eco-humanistic understanding and can lead to an unselfconscious engagement in the transformation of culture in the digital age. And in turn, this eco-humanistic understanding can guide us to use digital tools and techniques in a manner that embodies the analogue mode of cultural experience.

There are constant opportunities in the experiences of any society to begin looking at the familiar in new ways and to notice the unfamiliar that has been ignored. The challenge for social change agents is always to uncover the potential in the legacy of the familiar heritage, perceiving with diaphaneity and simultaneity, fully engaging in the cosmic dance between the desire for change and the comfort of hanging on to the existing.

The discussion above suggests the need for cunning intelligence, or what ancient Egyptians once conceived as the "intelligence of the heart". On the one hand, we can view digital technology as a means to an end, where the means and the end reciprocate purposefully in an integrative circularity. On the other hand, persevering and feeling comfortable with the paradoxical and tensional relationship between techne and teleos can bring forward a sustainable and desirable future that transforms the virtual digital reality into an authentic analogue experience and enhanced cultural reality. This is what I believe our digital age needs to consider for cultural transformation. Teleology is not the point of departure, nor is technology the destination of our making.

References

Ackoff, Russell L. *Redesigning the Future.* New York, NY: John Wiley, 1974. Print.

Debord, Guy. *Society of the Spectacle.* Trans. Ken Knabb. London: Rebel Press, n.d. Print.

Deely, John. *Semiotic Animal: A Postmodern Definition of "Human Being" Transcending Patriarchy and Feminism.* South Bend, IN: St. Augustine's Press, 2010. Print.

Detienne, Marcel, and Jean-Pierre Vernant. *Cunning Intelligence in Greek Culture and Society (European Philosophy and the Human Sciences).* Trans. Janet Lloyd. Chicago, IL: University of Chicago Press, 1991. Print.

Dunne, Joseph. *Back to the Rough Ground: Practical Judgment and the Lure of Technique.* Notre Dame, IN: University of Notre Dame Press, 1993. Print.

Gebser, Jean. *The Ever-Present Origin.* Trans. Noel Barstad with Algis Mickunas. Athens, OH: Ohio University Press, 1985. Print.

Grudin, Robert. *The Grace of Great Things: Creativity and Innovation.* New York, NY: Ticknor & Fields, 1990. Print.

Hall, Edward T. *Beyond Culture.* New York, NY: Anchor Books. 1976. Print.

Hawken, Paul. *Blessed Unrest, How the Largest Movement in the World Came into Being and Why No One Saw It Coming.* New York, NY: Viking Press, 2007. Print.

Heidegger, Martin. *The Question Concerning Technology and Other Essays.* Trans. William Lovitt. New York, NY: Harper and Row, 1977. Print.

Internet Encyclopedia of Philosophy. Web. 20 June 2010. http://www.iep.utm.edu/.

Jaki, Stanley L. *The Purpose of It All.* Washington, D.C.: Regnery Gateway. 1990. Print.

Jantsch, Erich. *Design for Evolution: Self-Organization and Planning in the Life of Human Systems.* New York, NY: George Braziller, 1975. Print.

Kanter, Rosabeth Moss. *Evolve!: Succeeding in the Digital Culture of Tomorrow.* Boston, MA: Harvard Business School Press, 2001. Print.

Lotman, Yuri M. *The Universe of the Mind: A Semiotic Theory of Culture.* Trans. Ann Shukman. Bloomington, IN: Indiana University Press, 1990. Print.

May, Rollo. *Love and Will.* New York, NY: Bantam Doubleday Dell Publishing Group, 1969. Print.

Mumford, Lewis. *The Conduct of Life.* New York, NY: Harcourt, Brace and Company, 1951. Print.

Peirce, Charles S. *Charles S. Peirce: Selected Writings (Values in a Universe of Chance).* Ed. Philip P. Wiener. New York, NY: Dover Publications, 1966. Print.

Pinto, Julio. 2009. "Can Semiotic Be the *Lingua Franca* for the Epistemological Hybrids of Contemporary Times?" *The American Journal of Semiotics.* 25.3-4 (2009): 67-73. Print.

Pirsig, Robert M. *Zen and the Art of Motorcycle Maintenance: An Inquiry into Values.* New York, NY: William Morrow and Company, 1979. Print.

Poster, Mark. *Information Please: Culture and Politics in the Age of Digital Machines.* London: Duke University Press, 2006. Print.

Ronchi, Alfredo M. *eCulture: Cultural Content in the Digital Age.* Dordrecht; Heidelberg; London; New York, NY: Springer, 2009. Print.

Seif, Farouk Y. "Communication in the Age of the Great Turning: Understanding the Role of Analog and Digital Modes in Liberating Imagination." *Communication: Understanding/ misunderstanding.* Proceedings of the 9th World Congress of the International Association for Semiotic Studies – Association Internationale de Sémiotique (IASS/AIS), Helsinki; Imatra: International Semotics Institute, 2007: 1626-1641. Print.

—. "Social Change in The 'Aperspectival World': The Paradox of Social Reproduction and Cultural Innovation." *TRANS. Internet-Zeitschrift für Kulturwissenschaften* 16(2005). Web. 10 Jan. 2010. http://www.inst.at/trans/16Nr/01_2/seif16.htm.

Farouk Y. Seif, PhD, is Professor Emeritus at Antioch University Seattle, USA where he teaches design for social/cultural change. He has taught, published, and presented his work internationally, urging reinterpretation of ancient wisdom for thoughtful reflections on contemporary life. Prof. Seif is the founder and director of Isis Institute where he leads the Intercultural Program in Egypt: Journey into the Dawn of Time.

www.ingramcontent.com/pod-product-compliance
Ingram Content Group UK Ltd.
Pitfield, Milton Keynes, MK11 3LW, UK
UKHW021848100926
8556IPUK00020B/1118

9 783631 640357